Modelling Investments in the World Copper Sector

MODELLING INVESTMENTS IN THE WORLD COPPER SECTOR

By Alfredo Dammert and Sethu Palaniappan

University of Texas Press, Austin

The authors are grateful to David Kendrick of the University of Texas at Austin, and to Ardy Stoutjesdijk, Alexander Meeraus and Kenji Takeuchi of the World Bank for their encouragement, support and comments during the various stages of the study.

International Standard Book Number 0-292-79026-0

Library of Congress Catalog Card Number 84-52196

Printed in the United States of America
First Edition, 1985

For reasons of economy and speed this volume has been printed from camera-ready copy furnished by the authors, who assume full responsibility for its contents.

Contents

Modelling Investments in the World Copper Sector

1.
Modelling the Copper Industry

1.1 Introduction

This study presents a programming approach to project design in the copper industry from a global perspective covering mining, processing and fabrication. Although the case of copper on a worldwide basis is considered here, a similar approach on a worldwide basis could be applied to other commodities. This chapter discusses the need for models of this type followed by a general description of such models. The last section presents a brief introduction to the contents of the book.

1.2 The Need for Planning Models: The Case of Copper

Industrial countries as well as developing countries have great interest in the exploitation of natural resources located in the latter. In order to sustain their own industrial production at minimum cost, industrial countries depend upon the natural resources of developing countries, either due to resource depletion in the former or because of the higher economic benefits of exploiting those in developing countries.

Developing countries rely heavily on their exports of natural resources in order to import equipment and intermediate inputs necessary for the development of their industrial basis. Therefore, it is of key importance for them to develop carefully planned policies with respect to their export capacity. Copper is among the most important non-ferrous metals, accounting for about twenty five percent by volume of the world's non-ferrous metals production. Developing countries account for about fifty percent of the world's primary copper production. The most important exporters among these countries are Chile, Zambia, Zaire and Peru. Copper provides about 50 percent of foreign exchange earnings for Chile, 90 percent for Zambia, 40 percent for Zaire and 14 percent for Peru. Besides its export as a raw material to earn foreign exchange, copper (as well as other non-ferrous metals) provides a potential economic advantage for establishing a semifabricating industry. The efficient development of the industry along these lines would provide the producing countries with additional export earnings as well as with a stronger industrial basis.

Investments in the copper industry are usually of considerable magnitude due to large infrastructure requirements such as railroad lines, ports, and power plants, and also due to economies of scale in the mining and processing stages. Table 1.1 indicates some recent investment plans in copper mining and processing in the world, indicating the magnitude of the financing required for such projects.

Because of the high investment requirements in the copper sector, developing countries depend heavily on foreign capital either in the form of direct investment, or as loans from international development and commercial banks. Thus, it becomes of crucial importance for developing countries to consider the viability of new copper projects, by examining their competitive position. Due to the international character of the copper industry, the selection and design of investment projects requires a global analysis of the world market.

1.3 The Use of Mixed Integer Linear Programming Models

The method used here to study alternative investments in copper mining, processing and semimanufacturing consists of a multiperiod linear mixed-integer programming model. The model is designed to minimize total discounted cost of investment, operation and transportation subject to (i) material balance constraints, (ii) capacity constraints which include investment activities, and (iii) market requirements. It permits explicit consideration of economies of scale in new investments.

The methodology followed in this study is based on that developed by Kendrick and Stoutjesdijk (1975) for industries which exhibit increasing returns to scale. The volume by Kendrick and Stoutjesdijk (1975) presents a methodology which incorporates the contributions by other authors on planning models with economies of scale. A brief survey of previous studies in this area will place in perspective the models presented in this study. Chenery (1952, 1959) developed quantitative methods to incorporate economies of scale into project planning, by considering in a first study the optimal timing of capacity construction for natural gas transmission followed by a study on interdependent projects that are subject to economies of scale. Vietorisz and Manne (1963) extended the methods initiated by Chenery and developed further by Markowitz and Manne (1957) to the spatial location of projects, by analyzing the optimal location of capacity in the South American fertilizer industry. Manne, et al. (1967) studied the optimal time-phasing, scaling and location of some industrial activities in India. Kendrick (1967) developed a sectoral multiperiod model of the Brazilian steel industry. A multi-sector model for South Korea was built by Westphal (1971) considering economies of scale in the steel and petroleum sectors.

More recent developments include a multi-country, multi-product dynamic model of the fertilizer industry by Stoutjeskijk, Frank and Meeraus (unpublished) for Kenya, Uganda and Tanzania. Later, Meeraus, Stoutjesdijk and Weigel (1975) built the first model on a worldwide basis which incorporates economies of scale. 1/ The models presented in this study are similar in structure to the latter but are adapted to the characteristics of the copper industry. For example, copper reserves classified according to ore grades are considered for each region.

With respect to the characteristics of these models, we may claim that their structure places in perspective the interrelationships among producing countries as well as between producers and markets. That is, the amount of copper produced in each country depends on the relative cost of production of that country with respect to other areas. Besides, a particular country (producer) may supply copper at an intermediate processing stage to another country for its further processing (blister to refineries or refined copper to semifabricators). Overall, the total amount produced must equal world market requirements.

The consideration of economies of scale in investment costs permits the evaluation of alternative investment strategies. For example, it permits one to determine the optimum number of plants to be built within a common market. The time dimension of the model permits the determination of the optimum timing of such projects.

The type of models presented here represent a necessary supplement to cost-benefit analysis. Cost-benefit analysis looks at single projects at a few specified locations assuming that certain important variables, such as markets are independent of projects at other locations. For example, one can imagine that the results of separate cost-benefit studies done by independent groups will be inconsistent with total market requirements. Therefore, decisions taken under such an approach may lead to inconsistencies in the data considered as given by the project analysts, thus invalidating the conclusions of their calculations. The approach presented in this study avoids such potential dangers by presenting the interrelationships among all the relevant economic units. It should be noted that this approach does not invalidate the need for more detailed studies of the solution to the model. A final argument with respect to the approach being considered in contrast with cost-benefit analysis, is that it provides a tool for project design as opposed to project selection. That is, the solution to the model specified here provides estimates of the optimal location, size and timing of new investments, while cost-benefit analysis consists of selecting among a portfolio of specific discrete projects.

1.4 Contents of the Study

The next chapter of the study presents a general view of the copper industry. Main producing and consuming areas are highlighted, and the technology of copper processing and fabricating is introduced. Cost estimates associated with each stage of copper processing are also included. An assessment of future demand for copper completes the chapter.

Chapter 3 describes the use of mixed integer linear programming models and their application to the copper industry. A model of the world copper industry is developed in detail. The world is divided into thirteen producing areas and thirteen markets. The stages of production are divided into: mining/milling, smelting, refining, and semifabricating (wire, tubes and sheet semifabricating plants). The model determines the patterns of investment required to satisfy future copper consumption requirements in the form of semimanufactures. The solutions to the model depend on copper reserves and their quality, production costs at each location, transport costs and for some versions, on import tariffs at each market. One set of scenarios considers a static version with 2000 as the target year. Another scenario includes time paths by considering investment patterns to fulfill demand in 1990, 1995 and 2000.

Chapter 4 shows the results of the model under different assumptions and the implications for producers and consumers. The scenarios include variations in costs of production as well as the effect of removing import tariffs. The main conlusions are presented in Chapter 5.

Table 1.1: COPPER MINE AND PLANT EXPANSION PLANS IN 1983

Company	Location	Project Type	Investment Capacity	US$ million	Start
Mexico de Cobre	Sonora, Mexico	Smelter/ refinery	180 tons tpy Cu	210	1985
Minera de Canada	Sonora, Mexico	Concentrator	50 tons tpd Cu concent.	250	1983
ASARCO	Hayden, Arizona, USA	Smelter	95 tons tpy Cu blister (modernization)	133	1984
Inspiration Consolidated	Globe-Miami, Arizona, USA	Smelter	145 tons tpy Cu blister (modernization)	450	1983/ 1993
Cia. Minera Aguilar	El Pachon,	Mine/Smelter	100 tons tpy Cu blister	1,200	Uncertain
Eluma S.A.	Carajas, Brazil	Mine/Smelter/ Refinery	150 tons tpy Cu	900	1987
Enami	Chile	Smelter expansion	From 160 to 320 tons tpy Cu blister	100	1986
Exxon	Disputada, Chile	Mine (open pit)	80 tons tpd Cu ore	1,200	1986
Utah de Chile	La Escondida, Chile	Complex	200 tons tpy Cu	1,500	1989
Rio Tinto Zinc/ Codemin	Cerro Colorado, Panama	Complex	90 tons tpd Cu ore	1,800	1990 (Uncertain)
Minero Peru	Cerro Verde, Peru	Mine (open pit) and concentrator	200 tons tpy Cu conc.	300	1985 (delayed)

Table 1.1: COPPER MINE AND PLANT EXPANSION PLANS IN 1983 (Continued)

Company	Location	Project Type	Investment Capacity	US$ million	Start
Minero Peru	Tintaya, Peru	Mine (open pit) and concentrator	160 tons tpy Cu conc.	327	1984
RTB Bor	Veliki Krivelj, Yugoslavia	Smelter	44 tons tpy Cu	366	1985
Zambia Consolidated Copper	Nchanga, Zambia	Plant	520 tons of Cu from stockpiled tailings for 15 years.	242	1984
Government	Saindak, Pakistan	Mine	12.5 tons tpd Cu ore	200	1984
Western Mining/ BP Australia	Roxby Downs, SA, Australia	Complex	150 tons tpy Cu	A 1,100	
Government	Sabah, Malaysia	Smelter	50 tons tpy Cu blister	152	
Consortium	OK Tedi, Papua, New Guinea	Complex	400 tons tpy concentrates	1,500	1986
Philippine Associated S&R (Pasar)	Leyte Island, Philippines	Smelter	138 tons tpy copper blister	41D	1983

tpd = metric tons per day.
tpy = metric tons per year.

Source: Engineering and Mining Journal, McGraw-Hill, New York, January 1983.

2.
The Supply of Copper and Its Uses

2.1 The Processes of Copper Production 2/

The organization of an industry depends to a large extent on the characteristics of its production processes. The description of the current technology in copper processing and semimanufacturing provides a framework for the analysis of the main aspects of the copper industry, such as the location of mines and plants, the investment needs for new capacity creation, and the structure of the market. Therefore, this section will cover the methods of production currently being applied in the copper sector.

The processes of copper production are illustrated in Fig. 2.1. These processes cover from mining to semimanufacturing. Copper is obtained from two sources: the mines which produce the so-called primary copper and recoveries from scrap which are called secondary copper and which account for about forty percent of the total copper consumption of the nonsocialist world.

Primary copper is obtained from mines located in industrial countries (mainly the US, Canada and the USSR) as well as in less-industrialized countries (the largest producers being Chile, Peru, Zambia and Zaire). Table 2.1 provides information on mine copper production for 1981. Since the cost of mining per ton of copper content is inversely proportional to the proportion of copper content of the ore, the latter factor is an important determinant for investment in copper mines.

Figure 2.1 Process of Copper Refining and Fabrication

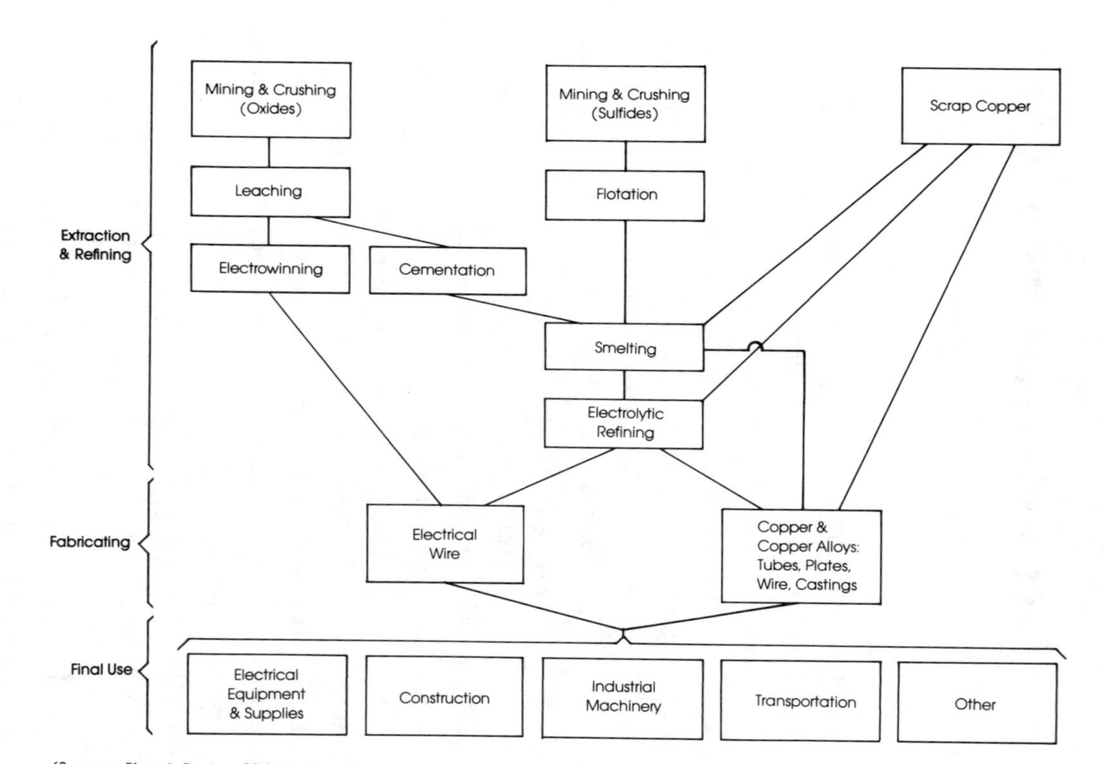

(Sources: Ffrench-Davis p. 92 & Martijena)

Table 2.1: MINE COPPER PRODUCTION AND CHARACTERISITCS, 1981

Countries	Production Cu Content 1,000 tons	Comments
Canada	718	Mostly nickel copper producers and copper zinc producers. High silver content.
United States	1,538	Ore grades 0.3-0.6%. Molybdenum as by-product.
Mexico and Central America	242	
Chile	1,081	Ore grade 0.8-1.2% Cu. Molybdenum as by-product.
Peru	328	Ore grade 0.8-1% Cu.
Western Europe	295	Ore grade 0.5-0.7% Cu.
Japan and Korea	52	Small mines. Ore grade about 0.4% Cu.
Papua New Guinea	165	About one half of the revenues come from gold and silver. Ore grade: 0.4% Cu, 0.6 grams gold/ ton of mineral.
Other Asia	451	Ore grade 0.4-0.5% Cu. High gold and silver content.
Australia	224	
Zaire	505	Open pit and underground mines. Complex ores. Cobalt and zinc by-products.
Zambia	587	Open pit and underground mines. Complex ores. Important cobalt by-products.
South Africa /1	306	
China	170	
USSR and E. Europe	1,648	
World Total	8,310	

/1 Includes Rest of Asia.

Sources: Production data: Metallgesellschaft, "Metal Statistics", Frankfurt am Main, 1981. Characteristics: World Bank.

The further processing of the ore depends on the type of mineral. Copper deposits can be broadly classified as copper oxides and copper sulfides. The copper oxides are soluble minerals. After milling the mineral is dissolved with sulfuric acid (leaching). The copper in solution may be recovered by precipitation with iron as cement copper (cementation), where the cement copper, averaging between 70% and 80% pure copper, is dewatered and shipped to a smelter. A second method consists in recovering copper from the leaching solution by an electrolytic deposition process complemented in some methods with a solvent extraction process (solvent extraction and electrowinning method). The product from this method requires no further processing prior to fabrication.

Copper sulfides are available throughout the world in much higher proportions than the oxides. Since sulfides are insoluble, they are concentrated by flotation by means of special reagents which capture the copper mineral. The resulting foams are separated and filtered in order to obtain copper concentrates whose copper content fluctuates between 20 and 50 percent according to the characteristics and composition of the mineral (French-Davis, 1974). In general, the concentration facilities are located near the mine due to the high transport costs that otherwise would take place.

The cement copper produced from copper oxides as well as the copper concentrates of the sulfide are processed at copper smelters where they are converted into blister copper which contains approximately 99 percent copper. Most smelting processes consist of three stages. The first step consists of treating the concentrate in a reverberatory furnace to eliminate part of the sulfur and to obtain molten copper matte. The molten copper is transferred to a converter where oxygen combines with the sulfur left in the matte in an exothermic reaction. Copper is removed from the converter at the end of the cycle with 99 percent purity and is referred to as blister copper. The one percent impurity includes gold, silver and other metals. Finally, the blister copper is subject to fire refining in order to oxidize and remove certain metals such as iron and nickel. Then, the copper is cast into 700 pound anodes (sheets of copper) and readied for shipment to an electrolytic refinery or to copper fabricating factories for which high purity is not required. In contrast with concentration, small mining operations do not typically have their own smelting facilities. This is due to the presence of economies of scale in smelters and also due to the fact that copper concentrates have a copper content from 50 to 100 times that of the mineral. This makes the transport cost per ton of copper content much lower than at the previous stage.

The blister or anode copper produced by the smelter is not of sufficient purity for many uses. Thus, after smelting, copper is shipped in most cases to an electrolytic refinery. In electrolytic refining the copper anodes are placed in cells (rectangular tanks) filled with an ionized solution (copper sulfate). When an electric current is applied to the cathodes and anodes, copper from the anodes flows to the thin sheets of copper which are the initial cathodes, until the latter gain weight to around 300 pounds. Each anode is utilized for 28 days and gives rise to two cathodes, one each 14 days. The impurities are collected on the bottom of the cells, containing, in some cases, significant amounts of gold and silver.

The copper semifabricating industry requires electrolytic copper in special shapes. The most common shapes are wirebar and copper billets of assorted sizes and shapes. To obtain these products the copper cathodes are melted in furnaces (reverberatory or other) from which the molten copper is sent to casting machines. This part of the process takes place in many cases at the site of the refinery, or in other cases the copper cathodes are sold directly to the semimanufacturing industries. Copper semifabricating or semimanufacturing can be divided into two main sectors: 1) fabrication of copper products for electrical uses such as electrical wire and cables, and 2) fabrication for other uses such as sheet, plate, rods, castings and non-electrical wire. About half the copper in the world is used for electrical conductors mainly in the form of wire produced from wirebars, first by a hot rolling process to produce wire rod and then by drawing. At various stages the metal has to be heated and then cooled, processed and annealed (U.N. Copper Production in Developing Countries, 1972). The production of non-electrical semimanufactures requires a different process. The fabrication starts by melting electrolytic copper (cathodes) in furnaces, in some cases together with blister copper, zinc or tin. To produce strip, sheet or plates the copper is molded into slabs, reheated, hot rolled and reheated and annealed at various stages. For rods, tubes and non-electrical wire, the molten copper is molded into billets and then it goes through different stages of machining, cutting, annealing and drawing according to the type of product required.

Finally, the different copper semimanufactures produced are shipped to sectors which consume copper. About half the copper in the world is used in electrical industries, fifteen percent in construction, fifteen percent in general engineering industries, ten percent in transportation and ten percent in other uses.

2.2 Location and Geology of Copper Deposits

Most copper deposits fall into one of five major types (Cox et al., 1973):

(1) The porphiry copper deposits account for about two-thirds of the world´s copper resources. These consist of disseminated copper sulfides in or near a felsic instrusive body. These type of deposits are located in the North American Southwest, British Columbia, the Philippines, Mexico and the Andean porphiry deposits of Chile and Peru. The principal byproducts of porphiry deposits are molybdenum, gold and silver.

(2) Strata bound deposits in sedimentary rocks, which include some of the world´s largest sources of copper. The origin of these deposits is uncertain. 4/ Main deposits of this type occur in Zambia and Zaire, US, (Michigan, Utah and New Mexico) and USSR (Udikan deposit). These deposits are the most important after the copper porphyries constituting one fourth of the world´s identifiable copper resources.

(3) About five percent of the copper resources are contained in volcanic rocks as massive sulfide deposits with an average of 2.5 percent copper. These deposits contain lead, silver and gold and are mainly located in Quebec (Noranda area), Japan and Cyprus.

Table 2.2: ESTIMATED WORLD COPPER RESERVES - 1981

(million tons of copper content)

Countries	Prices, 1981 US$/pound			
	0.75	1.00	1.25	1.50
Canada	3.4	9.1	9.3	16.6
United States	11.2	27.2	35.5	47.3
Mexico/Central America	-	3.6	12.8	24.1
Chile	13.6	55.8	64.8	72.2
Peru	-	1.7	1.7	14.7
Western Europe	5.8	7.4	7.5	8.2
Papua New Guinea	-	2.5	2.5	2.5
Philippines	0.7	4.9	6.9	8.9
Other Asia	3.3	5.4	5.6	11.3
Australia	0.6	6.4	6.4	6.8
Zaire	17.0	17.0	17.0	17.0
Zambia	14.2	18.0	19.0	19.0
South Africa	0.1	2.5	2.7	3.2
China	n.a.	n.a.	n.a.	n.a.
USSR/Eastern Europe	n.a.	n.a.	n.a.	n.a.
TOTAL (excluding China, USSR and E. Europe)	69.9	161.5	191.7	251.8

Reserves are cumulative: i.e., reserves at US$1/lb include those at US$0.75/lb, and so on.

Source: Computed from Barsotti, Aldo F. and Rodney D. Rosenkranz, "Natural Resources Forum," United Nations, New York, 1983.

(4) Copper as a byproduct in nickel ores. Significant producers of copper from these ores are located in Canada, Scandinavian countries, USSR and Western Australia.

(5) Native copper ores of the Keweenaw Peninsula (Michigan), Northwestern Canada and Chile.

Copper reserves have been increasing with time with the discovery of new deposits. Also, they vary with the copper price level. Table 2.2 shows the 1981 world copper reserves for prices from US$0.75/lb to US$1.50/lb of refined copper (in 1981 dollars).

2.3 Organization of the Industry

Financing investments in the copper industry requires considerable financial resources and knowhow. These requirements increased at the beginning of the 20th century with the industry shift towards large scale mining of relatively low ore grades. Thus, a handful of multinational enterprises, which possessed the necessary resources, expanded their activities around the world.

By 1960, the seven leading copper mining companies controlled 60% of the Western world's production of copper. However, this trend was reversed in the following two decades, with these companies' share of copper production decreasing to 23% by 1978 (see Table 2.3). According to Moran and Maddox, five factors were responsible for these developments: 4/

(a) A series of discoveries of new sources of copper;

(b) The diffusion of the large-scale mining technology;

(c) The availability of financing from copper processing plants which desired to secure raw material supplies (i.e., Japanese smelters);

Table 2.3: MINE CAPACITY OF COPPER PRODUCERS

(thousand metric tons)

	1960	1978
Kennecott	520	342
Anaconda	433	175
Anglo-American Group	356	153
Union Mineral	300	-
Roan-AMC Group	219	-
Phelps Dodge	213	315
INCO	141	151
CODELCO (Chile)	-	800
Gecamines (Zaire)	-	488
Zimco (Zambia)	-	390
ASARCO	-	326
Newmont	-	299

Sources: Moran, T.H. and D.H. Maddox, "Structure and Strategy in the International Copper Industry," UNCTC (draft), New York, 1980. Takeuchi, K., "Copper Handbook," World Bank, Washington, D.C., 1981.

(d) Diversification of other companies into copper mining;

(e) The emergence and growth of state-owned enterprises in several developing countries, which took over mining operations in Zaire (1967), Chile (1967-1971) and Zambia (1970).

Vertical integration is predominant in the copper industry. Many companies own mines, smelters, refineries as well as semifabricating facilities. The state mining companies in Zambia, Zaire, Chile and Peru have their own smelters and refineries. Among industrial countries, the US and Canada are the most important copper producers. Almost all smelting, refining and fabrication of their mine output takes place in these countries. ASARCO supplies its smelters and refineries in the US from mines situated in other countries, but which are partly owned by this company. Most North American and Japanese copper processing plants sell their output to their own semi-fabricating companies. Thus, in general, the supply of copper is still dominated by large producers.

2.4 Investment and Operating Costs

Both initial investment costs and operating costs of copper mining depend on different factors such as type of deposit, overburden thickness, haulage distance, ore hardness and the metallurgical qualities of the ore. The corresponding costs for smelting and refining depend on the copper content of the input as well as on its composition (Bennett, 1973). The main components of investment costs are: exploration costs, plant and equipment, physical infrastructure including roads, railways, power stations, water supplies, etc. and social infrastructure necessary for the provision and maintenance of the labor force, such as townships, schools, hospitals and training and recreational facilities (Prain, 1975). Table 2.4 presents estimates of investment costs for different regions of the world taking into account economies of scale. The differences in costs are due mainly to variations across countries in the factors mentioned above. In particular, investment costs in less industrialized countries tend to be higher due to larger requirements in infrastructure.

Operating costs vary widely for different mining methods and for different regions where costs of inputs and labor and the characteristics of the mineral may vary. Apart from these factors, by-products and co-products are important in determining operating costs. In some cases copper is a by-product as in the mines of the Sudbury Basin in Canada where nickel is the main product (Prain, 1971). Also, inflation and exchange rate movements have a considerable impact on the cost of producing primary copper.

Operating costs for primary copper production were estimated for different regions on the basis of unpublished data. Table 2.5 presents the operating costs after accounting for by-products.

The most important components of operating costs for mining and concentration are labor and supervision representing about 40 percent of the direct operating costs. For smelting (i) labor and (ii) fuel and power represent 50 and 25 percent of operating costs respectively. For refining, labor costs account for between 30 and 40 percent of direct operating costs while energy accounts for around 15 percent.

Let us now consider the costs of the production of copper semimanufactures. As mentioned earlier, about half of the

Table 2.4: INVESTMENT COSTS AND ECONOMIES OF SCALE

	Fixed Cost (FC)	Variable Cost (VC)	Maximum Size for Economies of Scale (MS)
	---(million 1980 US$)---		-('000 tons)-
Mines and Concentrates			
Developing countries (Zambia/Zaire)	120	5.8	380
Developing countries (all others) (ore grades 1% copper)	40	5.6	100
Industrial countries (ore grades 0.75% copper)	40	5.06	75
Smelters			
Developing countries	50	2.166	150
Industrial countries	40	1.832	150
Refineries			
Developing countries	5.02	0.718	150
Industrial countries	4.02	0.574	150

Source: Estimated from Engineering and Mining Journal (January Mine and Plant Expansion Survey, various issues), Bennett, Harold, et al., "An Economic Appraisal of the Supply of Copper From Primary Domestic Sources," Bureau of Mines Information Circular, U.S. Department of the Interior, Washington, D.C., 1973, and conversations with various companies.

Table 2.5: OPERATING COSTS

(1980 US$ per ton of copper content)

	Mining and Concentration (High Grade Ores)	Smelting	Refining
Peru, Mexico/C. America, South Africa	840	250	150
Chile	1,028	250	150
Zambia	1,044	250	150
Zaire	681	250	150
Philippines	1,054	250	150
Papua/New Guinea	295	250	150
Western/Eastern US	771	300	170
Other Industrial Countries	914	300	170

Source: Estimated from various projects and unpublished sources.

world's consumption of copper is in the form of electrical conductors, mainly electrical wire of different sizes with or without insulation. Most of the other half consists of copper or brass sheets, plates, tubes, rods and other shapes (Martijena, 1966; Prain, 1975). Investment costs in the production of copper semimanufactures are strongly affected by economies of scale. Table 2.6 presents investment costs for the main lines of copper semifabricates and for different plant sizes.

It can be seen from the data presented above that economies of scale exert an important influence in the capital requirements of copper semimanufactures. Thus the investment cost per ton for a plant which produces 10,000 tons per year of electrical wire amounts to 75% of that of a 3,000 tons per year plant; the cost per ton for plants which produce tubes and rods or sheets, plate and strip falls to 46% of the smaller plant cost as the size of the plant increases from 3,000 to 20,000 tons per year.

Operating costs for copper semimanufactures depend on the product mix both with respect to the types of products and its diversity. Besides, the cost of labor has a significant impact on operating costs. Table 2.7 presents costs of production for electrical wire, tubes and rods and sheet, plate and strip for different regions of the world without the cost of refined copper. The difference in costs reflects the wage differences between regions. Table 2.8 presents the product mix on which the calculations are based.

From Table 2.7 we may observe that due to wage differentials and assuming the same technology and plant size, the cost of processing copper into electrical wire is about 40% lower in China and Other Asia, and about 30% lower

Table 2.6: INVESTMENT COSTS FOR COPPER SEMIMANUFACTURING PLANTS

	Fixed Cost (FC)	Variable Cost (VC)	Maximum Size for Economies of Scale (MS)
	(million 1980 US$)	(million 1980 US$/'000 tons)	('000 tons)
Wire			
Developing countries	3.6	1.84	20
Industrial countries	3.38	1.72	20
Tubes and Rods			
Developing countries	6.0	1.0	30
Industrial countries	5.64	0.94	30
Sheets, Plates and Strip			
Developing countries	9.2	1.44	30
Industrial countries	8.64	1.34	30

Source: Updated from Dammert, Alfredo, "A World Copper Model for Project Design", (unpublished Ph.D. dissertation) The University of Texas at Austin, Austin, 1977.

Table 2.7: ESTIMATES OF OPERATING COSTS FOR COPPER SEMIMANUFACTURES (excluding the cost of copper)

(1980 US$/ton)

	Wire	Bars and Tubes	Sheet Plate and Strip
Mexico, South America, Central Africa and South Africa	1,330	1,507	2,090
China and Other Asia	1,150	1,447	2,000
Industrial Countries	1,870	1,687	2,360

Source: Updated from Dammert, Alfredo, "A World Copper Model for Project Design", (unpublished Ph.D. dissertation), The University of Texas at Austin, Austin, 1977.

Table 2.8: PRODUCT MIX FOR COPPER SEMIMANUFACTURES

	Percentage (Gross Tonnage)
Electrical Wire	
1.6 mm diameter copper wire	13.73
7 x 1.05 mm diameter copper wire	1.72
7 x 0.85 mm diameter rubber insulated	26.07
1.6 mm diameter PVC insulated	56.86
7 x 1.05 mm diameter PVC insulated	1.59
Bars and Tubes	
Copper bars and profiles	11.66
Brass bars and profiles	46.64
Copper tubes	8.34
Brass tubes	33.36
Plate and Strip	
Copper sheet and strip	15.0
Brass sheet and strip	85.0

Source: Updated from Dammert, Alfredo, "A World Copper Model for Project Design" (unpublished Ph.D. dissertation), The University of Texas at Austin, Austin, 1977.

in other developing countries when compared with industrial countries. This differential is much narrower for the other copper semimanufactures, with processing costs about 15% lower in China and Other Asia, and 11% lower in other developing countries with respect to industrial countries.

2.5 Transportation Costs and Tariffs

Transportation costs and tariffs play a significant role in the location of processing plants as well as in shipment patterns both for copper and for its semimanufactures. In later chapters, the importance of these factors will be evaluated. This section outlines their main characteristics.

Due to the fact that the copper content of copper ores range from 0.4% to 3%, concentrators are usually located near the mines (Bennett, 1973). Copper concentrates range between 20% and 40% of copper content which makes the freight rates per ton of copper content about three times that of blister or refined copper, which have 99% or more pure copper content. Despite this fact some industrialized countries (mainly Japan, Germany and the US) import about 11% of copper concentrates either from subsidiary companies or from firms whose financing is dependent on the importers (Banks, 1974).

Since the difference between the copper content of blister and that of refined copper is of less than one percent, their transportation costs per ton of copper content are virtually the same. Therefore transportation costs have little effect on the location of copper refineries as long as they are located along the route from the smelter to the major copper markets. This is illustrated by the fact that in 1969, before major nationalizations, less industrialized countries refined only about 64 percent of their mine production, while in 1974 less industrialized copper-producing countries refined 83% of their own mine production. With respect to actual transportation patterns, Banks (1974) suggests that the optimal trade pattern is being distorted by ownership ties, political considerations and the like. Due to higher bulk and storage difficulties, transportation costs of copper semimanufactures are approximately twice those of refined copper as shown in Table 2.9.

Besides higher freight rates for semimanufactures, industrialized countries have placed discriminatory tariffs against these products in order to protect their local industry (Baranson, no date; Banks, 1974). Table 2.10 presents the current tariff structure on copper products. Thus, the existence of higher freight rates and tariffs makes it more difficult for less industrialized countries to establish copper semimanufacturing industries for export markets.

2.6 Secondary Copper

As mentioned previously, scrap plays a vital role in the supply of copper. New scrap, besides being recycled by fabricators, comes from waste arising from manufacturers and end users, while old scrap consists of material salvaged from obsolete equipment and machinery (Prain, 1975). The better qualities of new scrap are often used directly by fabricators of non-electrical semimanufactures. Lower quality scrap goes to smelting or directly to refining according to its quality (Prain, 1975).

Table 2.9: AVERAGE TRANSPORTATION COSTS OF REFINED COPPER AND SEMIMANUFACTURES

(1980)

	Copper Concentrates	Blister and Refined Copper	Copper Semimanufactures
Ocean Transport			
Loading and Unloading US$/ton	3.00	4.00	8.00
Shipment US$/ton-nautical mile	.007	.01	.02
Railroad Transport			
Tariff US$/ton-mile	.04	.05	.1

Sources: Dammert, Alfredo, "Economic Minera," Universidad del Pacifico, Lima, 1980.

Consultations with shipping companies.

Table 2.10: AD VALOREM DUTIES ON COPPER

(percentages)

	Copper Metal	Semimanufactured Copper Products
E.C.	0	8
USA	1	6
Canada	0	6
Japan	8.5	15

Sources: For copper metal--Takeuchi, Kenji, "Copper Handbook," World Bank, Washington, D.C., 1981.

For Semimanufactures--UNCTAD, "Processing and Marketing of Copper: Areas for International Cooperation," Geneva, 1982.

Table 2.11 shows the quality of scrap available at smelters, refineries and semimanufacturers (net of recycling) for 1980.

The supply of new scrap increases proportionally to current consumption due to the fact that it is collected from the copper manufacturers. The collection of old scrap is much more complex, depending directly on the price of refined copper. This is related to the higher cost of collection of old scrap and its higher content of impurities.

2.7 Uses and Consumption of Copper

Copper has a wide variety of applications due to its many important characteristics. Prain 5/ cites its most important properties as: its high electrical and thermal conductivity; ease of working and forming; ability to be alloyed with many other metals permitting a wide range of properties; ease of jointing by soldering and brazing; good mechanical properties such as tensile strength, elongation and hardness; high resistance to many corrosive environments; good electro-deposition characteristics and suitability for architectural, artistic and decorative purposes. In accordance with these properties, the main sectors which use copper are electricity, construction, transportation, general engineering, and consumer goods. Table 2.12 presents the consumption breakdown according to sectoral uses for the US, Western Europe and Japan.

If we take the countries which appear in Table 2.12 as representative, we may conclude that the electrical engineering industry together with the general engineering industry, the construction industry and the transportation industry, account for about 90% of the world's consumption

Table 2.11: COPPER SCRAP AVAILABILITY, 1980

('000 of metric tons)

	Smelters	Refineries	Semi-manufacturing Plants
Canada	-	40	19
Australia	5	28	43
Western Europe	130	460	920
Japan/Korea	50	141	400
USA	60	440	880
Brazil	-	27	25
Mexico	-	5	-

Source: Metal Statistics, Metalgesselschaft, Frankfurt Am Main, 1981.

Table 2.12: CONSUMPTION OF COPPER BY END USES, 1981

(%)

	Electrical	Construction	Transportation	General Engineering	Consumer Goods and Other
USA	50	15	7	20	8
Western Europe	48	15	8	21	8
Japan	51	8	16	18	7

Source: Computed from Brook Hunt & Associates, Ltd., Quarterly Service, Issue XII, Copper, London, December 1982.

of copper. Due to the high conductivity of copper, about one half of the world's consumption is by the electrical industry, mainly in the form of electrical wires and cables, windings of motors and transformers and cables for the communications industry. In certain uses of the electrical industry, copper faces the competition of aluminum, as in the case of underground power cables in the lower voltage ranges. In the communications industry the intensity of copper use is diminishing due both to copper saving innovations (thinner wires, improved efficiency) and by replacement of copper wires by optical fiber cables. A significant portion of copper cables in telecommunications will be replaced by fiber optics in the next two decades. Their advantages are many: lower cost, smaller size and better technical properties.

In the construction sector copper faces strong competition from various materials. Copper tubes and fittings for water services account for about one-half of copper uses in this sector. Substitution is stronger in Western Europe and Japan than in the United States, as in the first two areas black iron, lead and zinc pipes represent alternatives to copper pipes.

Copper consumption in transportation depends mainly on the growth rate of automobile output and of transport equipment. Copper saving practices in automobile manufacturing as well as substitution by aluminum windings in dynamos and starter motors and steel radiator fins affect the growth of copper consumption in these areas. The main threat, however, is in car radiators. Aluminum car radiators were introduced in US cars in 1981 and in Japanese cars in 1982. 6/ In shipbuilding and railways there is no present trend for substituting copper by other metals, but these subsectors are likely to experience lower growth than in past decades.

In general engineering the most important uses are for valves, tubes, pump bearings and similar products which account for about 45% and industrial heat exchangers which use about 20% of the consumption of copper in that sector. In valves and bearings there has been substitution of copper by other materials such as stainless steel and aluminum. In refrigeration, aluminum is increasingly being used. In tubing, alternative materials are steel and titanium which have longer life although they have also higher fabricating and machining costs.

Turning to the main shapes of copper semimanufactures, these are: wires and cables; tubes and pipes; sheet, plate and strip; and others. For this study, an attempt has been made to estimate total consumption of copper semimanufactures by shapes and regions. For main consuming countries, this breakdown was available. For other countries or regions the following formula was used.

$$C_s = f \times T + I_s - E_s$$

where
- C_s: Consumption of semimanufactures
- T: Total use of copper
- f: Factor which relates the total use of copper to production of semimanufactures
- I_s: Imports of semimanufactures
- E_s: Exports of semimanufactures

After obtaining estimates of total consumption of copper semimanufactures for the latter group of countries, the same distribution by shapes as for main consuming countries was employed. The results are shown in Table 2.13, including a forecast which is explained below.

Table 2.13: CONSUMPTION OF COPPER SEMIMANUFACTURES 1980-2000 ('000 metric tons)

	1980					1990				
	Wire	Tubes	Sheet	Other	Total	Wire	Tubes	Sheet	Other	Total
United States	1,385	690	555	40	2,670	1,647	821	660	48	3,176
Canada	125	60	50	5	240	150	72	60	6	288
Mexico	80	40	30	4	154	118	59	44	6	227
Eastern South America	195	95	80	16	386	287	140	118	24	569
Western South America	40	20	15	3	78	59	29	22	4	114
Western Europe	2,075	1,160	500	205	3,940	2,432	1,359	586	240	4,617
Central Africa	15	7	5	1	28	22	10	7	1	40
South Africa	55	30	20	3	108	81	44	29	4	158
Japan and Korea	875	440	350	45	1,710	1,240	624	496	64	2,424
China	195	100	80	11	386	258	132	106	15	511
Other Asia	90	45	35	6	176	133	66	52	9	260
Australia	85	40	35	6	166	102	48	42	7	199
USSR/E. Europe	1,050	525	420	55	2,050	1,404	702	562	74	2,742
TOTAL	6,265	3,252	2,175	400	12,092	7,933	4,106	2,784	502	15,325

	2000				
	Wire	Tubes	Sheet	Other	Total
United States	2,048	1,020	821	59	3,948
Canada	190	91	76	8	365
Mexico	174	87	65	9	335
Eastern South America	425	207	174	35	841
Western South America	87	44	33	7	171
Western Europe	3,023	1,690	728	299	5,740
Central Africa	33	15	11	2	61
South Africa	120	65	44	7	236
Japan and Korea	1,784	897	713	92	3,486
China	327	168	134	18	647
Other Asia	196	98	76	13	383
Australia	129	61	53	9	252
USSR/E. Europe	1,887	944	755	99	3,685
TOTAL	10,423	5,387	3,683	657	20,150

Sources: For 1980 - Metallgesellschaft "Metal Statistics," Frankfurt am Main, 1981; SAMIM, "Yearbook".
Projections - World Bank estimates.

For 1990 and 2000, estimates of copper semimanufactures consumption were made using as a basis, World Bank's projections for copper consumption. Taken as a whole, world consumption of copper semimanufactures is expected to grow at an average of 2.6% per year from 1980 to 2000. The main consuming regions will continue to be the United States, Western Europe, Japan and USSR/Eastern Europe, accounting for about 85% of total world consumption. Among developing regions, Eastern South America and China represent the most important markets for copper semimanufactures with 841,000 tpy and 647,000 tpy respectively by 2000, which represents about 7% of world consumption. In general, the figures show that the main copper markets would still remain in the developed countries and that the strategies for copper producing countries should be aimed at reaching the former markets with higher value added copper products.

3.

Characteristics of the Model

3.1 Model Description

The model is specified as a mixed integer problem required to minimize total systems cost. The objective function is therefore the sum of transport costs, operating costs, tariff costs and investment charges (annualized rentals) over time. This minimization is subject to satisfying material balances, and capacity and market requirements. The "mixed-integer" component arises from the need to build new capacity at various locations to satisfy increased demand. The number of locational choices and sizes that exist subject to trade offs between investment charges, operating costs and transport costs makes this a nontrivial problem.

The copper industry in the model is segregated into four vertical components, ore production (mines), blister smelting (smelters), copper refining (refineries), and copper semi-manufacturing (wire fabrication, tubes and rods, and sheets, plates and strip). A key characteristics of the model is the classification of ore grades into high grade and medium grade ores, to account for quality and extraction cost differences. A set of 15 geographical aggregations define the locations that have either existing capacity or could build new capacity or add to existing capacity in the copper processing subsectors (mines, smelters and refineries). These 15 locations are further distinguished by their ability to process copper scrap inputs. A set of

13 geographical aggregations define both the copper semi-manufacturing locations and markets. Both these geographical aggregations were determined on the basis of data availabilities (and quality), geographical structures, industrialization levels, and computational restriction on problem size.

The base run and experiments were performed on a single planning period, for the year 2000. The problem then becomes one of examining what the industry requirements would be in the year 2000, given only the industry structure in 1980 and demand projections for the year 2000. A final experiment, using the assumptions of the base run, was run over multiperiods. This provided more detail on the time-phasing of the capacity additions in the system for a specific scenario.

3.2 Model Specification

3.2.1 Sets

$i, i' \in I$ — Mine, refinery and smelter (processing) locations

Peru	
Chile	
Zambia	
Zaire	
Mex+Cam	Mexico and Central America
S-Africa	South Africa
Philippines	
Papua-Ng	Papua-New Guinea
Western-US	Western United States
Eastern-US	Eastern United States
Canada	

	EE+USSR	Eastern Europe and Soviet Union
	Australia	
	W-Europe	Western Europe
	Japan+Kor	Japan and South Korea
$j, j' \varepsilon J$	Wire, tubes and rods, and sheet (semi-manufacturing) plant locations	
	USA	
	Mex+Cam	Mexico and Central America
	ES-America	Eastern South America
	WS-America	Western South America
	W-Europe	Western Europe
	EE+USSR	Eastern Europe and Soviet Union
	C-Africa	Central Africa
	S-Africa	South Africa
	O-Asia	Other Asia
	Japan+Kor	Japan and South Korea
	China	
	Australia	
	Canada	
$c \varepsilon C$	Commodities	
	ore	
	scrap-s	scrap used in smelting
	blister	
	scrap-r	scrap used in refining
	refined-cu	refined copper
	scrap-sps	scrap used in sheet, plate and strip manufacture

	sheets+p+s	sheets, plates and strips
		wire
	scrap-t	scrap used in tube and rod manufacture
	tubes+rods	tubes and rods

$c \in CM$ Commodities at mines, refineries and smelters

ore
scrap-s
blister
scrap-r
refined-cu

$c \in CS$ Commodities at semi-manufacturing plants

refined-cu
wire
scrap-t
tubes+rods
scrap-sps
sheets+p+s

$c \in CF$ Final products

refined copper
wire
tubes+rods
sheets+p+s

c ε CFS	Final products at semi-manufacturing plants	
	wire	
	tubes+rods	
	sheets+p+s	
c ε CIM	Intermediate products at refineries and smelters	
	ore	
	blister	
c ε CIL	Scrap inputs	
	scrap-s	
	scrap-r	
	scrap-t	
	scrap-sps	
p ε P	Processes	
	high-grade	high grade ore mining by open-pit and concentrator
	med-grade	medium grade ore mining by open-pit and concentrator
	smelting	
	smelt-s	smelting using scrap input
	refining	
	ref-s	refining using scrap input
	wire-ref-c	wire fabrication using refined copper
	tube-ref-c	tube and rod semi-manufacture using refined copper

	tube-scrap	tube and rod semi-manufacture using scrap
	s-ref-c	sheet semi-manufacture using refined copper
	s-scrap	sheet semi-manufacture using scrap
$p \varepsilon PM$	Processes at mines, refineries and smelters	
	high-grade	
	med-grade	
	smelting	
	smelt-s	
	refining	
	ref-s	
$p \varepsilon PSM$	Processes at semi-manufacturing plants	
	wire-ref-c	
	tube-ref-c	
	tube-scrap	
	s-ref-c	
	s-scrap	
$p \varepsilon PMM$	Mining processes	
	high-grade	
	med-grade	
$m \varepsilon M$	Productive units	
	open-pit	mines with concentrators
	smelter	
	refinery	
	wire	fabrication unit

	tubes+rods	tube and rod manufacturing unit
	sheets+p+s	sheet, plate and strip unit

$m \in MM$ — Productive units at mines, refineries and smelters

open-pit

smelter

refinery

$m \in MS$ — Productive units at semi-manufacturing plants

wire

tubes+rods

sheets+p+s

$t, t' \in T$ — Time periods

3.2.2 Variables

$Z^{m}_{p,i,t}$ — process levels at mines, refineries and smelters

$Z^{s}_{p,j,t}$ — process levels at wire, tube and sheet plants

$X^{i}_{c,i,i',t}$ — interplant shipments of ore and blister

$X^{fr}_{i,j,t}$ — shipments of refined copper to markets

$X^{ir}_{i,j,t}$	shipments of refined copper to semimanufacturing plants
$X^{fs}_{c,j,j',t}$	shipments of semimanufactures to markets
$A^{m}_{c,i,t}$	scrap availability for smelting and refining
$A^{s}_{c,j,t}$	scrap availability for tube and sheet manufacture
$A^{ss}_{j,t}$	scrap availability for semimanufacturing
$H^{m}_{m,i,t}$	capacity expansion at mines, refineries and smelters
$H^{s}_{m,j,t}$	capacity expansion at wire, tube and sheet plants
$S^{m}_{m,i,t}$	unused economies-of-scale expansion at mines, refineries and smelters
$S^{s}_{m,j,t}$	unused economies-of-scale expansion at wire, tube and sheet plants
$Y^{m}_{m,i,t}$	expansion decision variable at mines, refineries and smelters
$Y^{s}_{m,j,t}$	expansion decision variables at wire, tube and sheet plants

Φ_t^{rm}	annualized rental charges for mines, refineries and smelters
Φ_t^{ks}	annualized rental charges for wire, tube and sheet plants
Φ_t^{om}	operating costs for mines, refineries and smelters
Φ_t^{os}	operating costs for wire, tube and sheet plants
Φ_t^{tr}	transport costs
Φ_t^{tf}	tariffs
Φ	total cost

3.2.3 Parameters

$a_{c,p}$	input-output matrix
$b_{m,p}$	capacity utilization
$d_{j,c,t}$	final demand
$r_{i,p}$	ore reserves at mines
$a^m_{c,i,t}$	scrap available for smelting and refining

$a^{s}_{c,j,t}$ scrap available for tube and sheet manufacture

$a^{ss}_{j,t}$ scrap available for semimanufacturing

$k^{m}_{i,m}$ initial capacity of mines, smelters and refineries

$k^{s}_{j,m}$ initial capacity of wire, tube and sheet plants

$\hat{h}^{m}_{i,m}$ maximum size of mines, smelters and refineries

$\hat{h}^{s}_{j,m}$ maximum size of wire, tube and sheet plants

$\bar{h}^{m}_{i,m}$ economies-of-scale size of mines, smelters and refineries

$\bar{h}^{s}_{j,m}$ economies-of-scale size of wire, tube and sheet plants

ξ useful life of productive units

σ capital recovery factor

s^{fm}_{i} site factor for mines, smelters and refineries

s^{fs}_{j} site factor for wire, tube and sheet plants

$\omega^{m}_{i,m}$	scale cost of mines, smelters and refineries
$\omega^{s}_{j,m}$	scale cost of wire, tube and sheet plants
$\nu^{m}_{i,m}$	proportional capital cost of mines, smelters and refineries
$\nu^{s}_{j,m}$	proportional capital cost of wire, tube and sheet plants
ζ^{pi}_{i}	operating efficiency factor of mines, smelters and refineries
ζ^{pi}_{i}	operating efficiency factor of wire, tube and sheet plants
$o^{pm}_{i,pm}$	operating cost for mines, refineries and smelters
$o^{ps}_{j,psm}$	operating cost for wire, tube and sheet plants
$\mu^{r}_{i,i',c}$	transport costs for ore and blister
$\mu^{i}_{i,j}$	transport costs for refined copper
$\mu^{fs}_{j,j}$	transport costs for wire, tubes and sheets
$t^{r}_{i,j}$	tariffs on refined copper

$t^s_{j,j}$ tariffs on wire, tubes and sheets

n_t length of model horizon

δ_t discount factor

3.2.4 Equations

Material balance at mines, smelters and refineries

$$\sum_{\substack{p \in PM \\ (i,p) \in MAPIP}} a_{c,p} \cdot Z^{m}_{p,i,t} \quad + \sum_{i' \in I} X^{i}_{c,i',i,t} \quad \Big|_{c \in CIM}$$

[process requirements (-) and output (+) of commodity c at location i, only for processes allowed at i] + [interplant shipments of ore or blister coming into i from all other locations i′]

$$+ \quad A^{m}_{c,i,t} \quad \Big|_{c \in CIL}$$

+ scrap availability at location i

$$\geq \sum_{j \in J} X^{fr}_{i,j,t} \Big|_{c \,\in\, \text{refined copper}} + \sum_{i' \in I} X^{i}_{c,i,i',t} \Big|_{c \in CIM}$$

$$\geq \left[\begin{array}{l} \text{shipments of refined copper} \\ \text{as end product to all markets} \\ \text{from i} \end{array} \right] + \left[\begin{array}{l} \text{interplant shipments of ore} \\ \text{and blister going out of i} \\ \text{to all other locations i'} \end{array} \right]$$

$$+ \sum_{j \in J} X^{ir}_{i,j,t} \Big|_{c \in \text{refined copper}} \qquad \begin{array}{r} c \in CM \\ i \in I \\ t \in T \\ (i,c) \in MAPIC \end{array}$$

$$+ \left[\begin{array}{l} \text{shipments of refined copper for intermediate} \\ \text{use from i to all semi-manufacturing locations} \end{array} \right]$$

<u>Material balance at wire, tube and sheet plants</u>

$$\sum_{\substack{p \in PSM \\ (j,p) \in MAPJP}} a_{c,p} \cdot Z^{s}_{p,j,t} + \sum_{i \in I} X^{ir}_{i,j,t} \quad c \in \text{refined copper}$$

$$\left[\begin{array}{l} \text{process requirements (-) and} \\ \text{output (+) of commodity c at j} \end{array} \right] + \left[\begin{array}{l} \text{shipments of refined copper for} \\ \text{use as intermediate input from} \\ \text{all refineries to location j} \end{array} \right]$$

$$+ \; A^{s}_{c,j,t} \Big|_{c \varepsilon CIL}$$

$$+ \left[\begin{array}{l} \text{scrap availability for copper} \\ \text{semi-manufacturing at } j \end{array} \right]$$

$$\geq \sum_{j' \varepsilon J} X^{fs}_{c,j,j',t} \Big|_{c \varepsilon CFS} \qquad \begin{array}{l} c \; \varepsilon \; CS \\ j \; \varepsilon \; J \\ t \; \varepsilon \; J \\ (cmj) \; \varepsilon \; MAPJC \end{array}$$

$$\geq \left[\begin{array}{l} \text{shipments of copper semi-manufacture} \\ \quad c \text{ from } j \text{ to all markets } j' \end{array} \right]$$

Market Requirements

$$\sum_{i \varepsilon I} X^{fr}_{i,j,t} \Big|_{c \varepsilon \text{ refined copper}} + \sum_{j' \varepsilon J} X^{fs}_{c,j',j,t} \qquad \begin{array}{l} j \ \varepsilon \ J \\ c \ \varepsilon \ CF \\ t \ \varepsilon \ T \end{array}$$

[shipments of refined copper as end product from all refineries i to market j] + [shipments of finished copper semi-manufacture c from all semi-manufacturing locations j′ to market j]

$$\geqslant \quad d_{j,c,t}$$

$\geqslant$ [demand for final products at market j]

Capacity constraints at mines, smelters and refineries

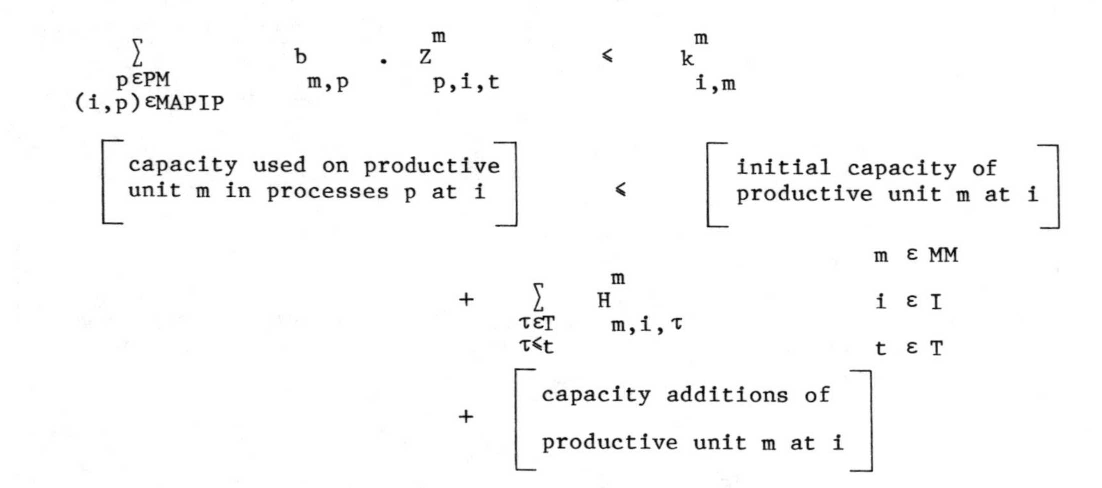

$$\sum_{\substack{p \varepsilon PM \\ (i,p) \varepsilon MAPIP}} b_{m,p} \cdot Z^{m}_{p,i,t} \leqslant k^{m}_{i,m} + \sum_{\substack{\tau \varepsilon T \\ \tau \leqslant t}} H^{m}_{m,i,\tau} \qquad m \ \varepsilon \ MM, \quad i \ \varepsilon \ I, \quad t \ \varepsilon \ T$$

$$\left[\begin{array}{l}\text{capacity used on productive} \\ \text{unit m in processes p at i}\end{array}\right] \leqslant \left[\begin{array}{l}\text{initial capacity of} \\ \text{productive unit m at i}\end{array}\right] + \left[\begin{array}{l}\text{capacity additions of} \\ \text{productive unit m at i}\end{array}\right]$$

Capacity constraints at wire, tube and sheet plants

$$\sum_{\substack{p \in PSM \\ (j,p) \in MAPJP}} b_{m,p} \cdot Z^{s}_{p,j,t} \leq k^{s}_{j,m} + \sum_{\substack{\tau \in T \\ \tau \leq t}} H^{s}_{m,j,\tau} \qquad m \in MS, \; j \in J, \; t \in T$$

[capacity used on productive unit m by processes p at location j] ≤ [initial capacity of productive unit m at location j] + [capacity additions of productive unit m at location j]

Maximum expansion at mines, smelters and refineries

$$H^{m}_{m,i,t} \leqslant \hat{h}^{m}_{i,m} \cdot Y^{m}_{m,i,t} \qquad m \in MM,\; i \in I,\; t \in T$$

$$\begin{bmatrix}\text{capacity expansion of}\\ \text{productive unit m at}\\ \text{location i in period t}\end{bmatrix} + \begin{bmatrix}\text{maximum expansion of}\\ \text{productive unit m}\\ \text{allowed at location i}\end{bmatrix}$$

Maximum expansion at wire, tube and sheet plants

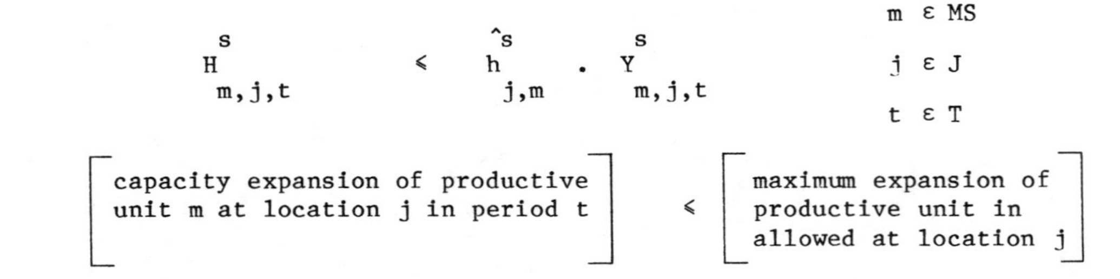

$$H^{s}_{m,j,t} \leqslant \hat{h}^{s}_{j,m} \cdot Y^{s}_{m,j,t} \qquad m \in MS,\; j \in J,\; t \in T$$

$$\begin{bmatrix}\text{capacity expansion of productive}\\ \text{unit m at location j in period t}\end{bmatrix} \leqslant \begin{bmatrix}\text{maximum expansion of}\\ \text{productive unit in}\\ \text{allowed at location j}\end{bmatrix}$$

Limits to economies-of-scale constraint for mines, smelters and refineries

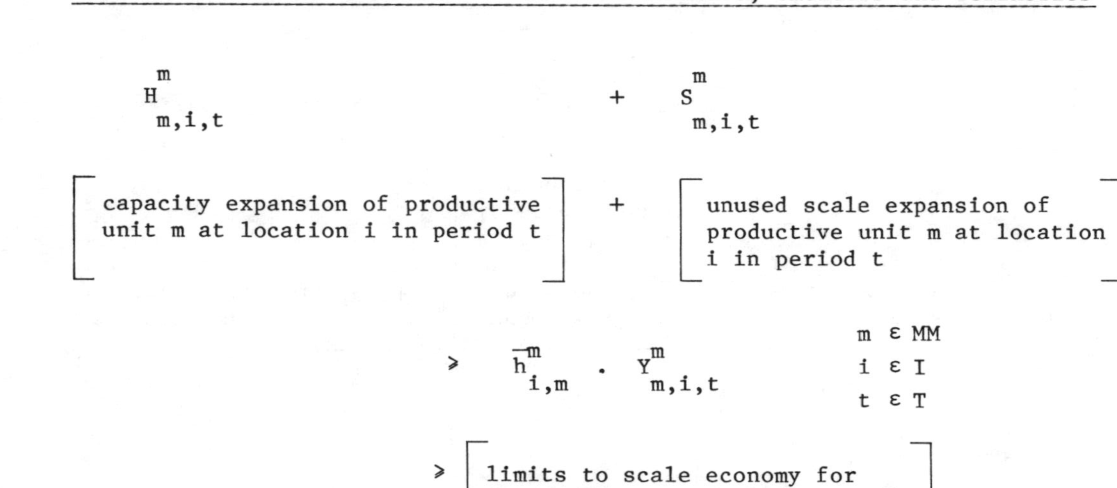

$$H^{m}_{m,i,t} + S^{m}_{m,i,t}$$

[capacity expansion of productive unit m at location i in period t] + [unused scale expansion of productive unit m at location i in period t]

$$\geq \bar{h}^{m}_{i,m} \cdot Y^{m}_{m,i,t} \qquad m \in MM,\; i \in I,\; t \in T$$

$\geq$ [limits to scale economy for productive unit m at location i]

Limits to economies-of-scale constraint for wire, tube and sheet plants

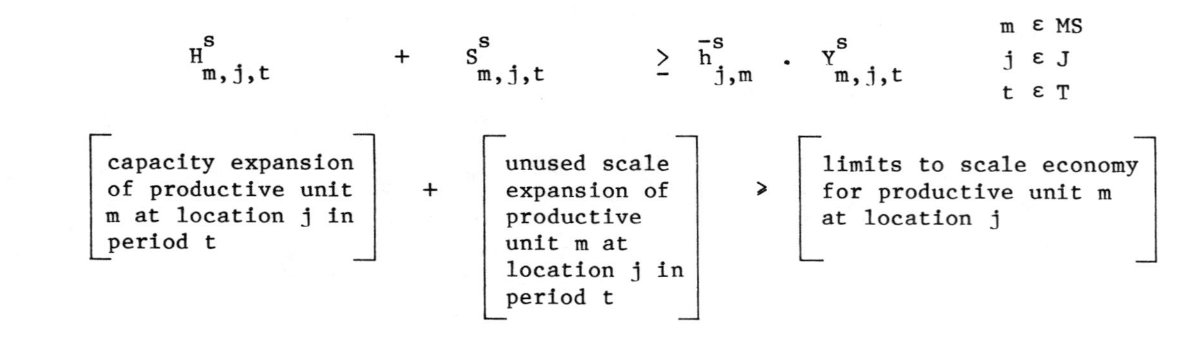

$$H^{s}_{m,j,t} + S^{s}_{m,j,t} \geq \bar{h}^{s}_{j,m} \cdot Y^{s}_{m,j,t} \qquad \begin{array}{l} m \in MS \\ j \in J \\ t \in T \end{array}$$

[capacity expansion of productive unit m at location j in period t] + [unused scale expansion of productive unit m at location j in period t] > [limits to scale economy for productive unit m at location j]

Scrap aggregation at semi-manufacturing plants

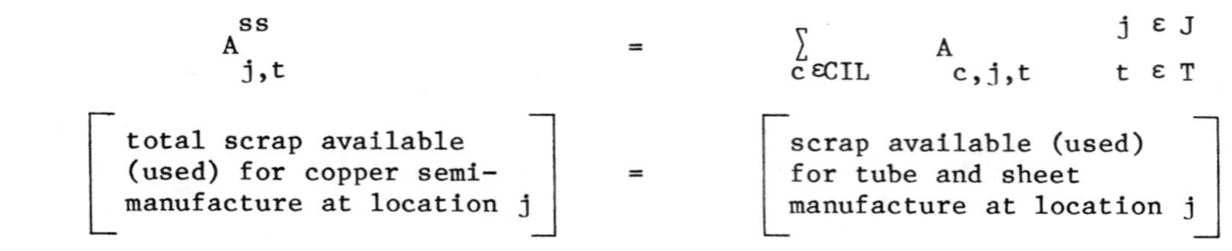

$$A^{ss}_{j,t} = \sum_{c \in CIL} A_{c,j,t} \qquad \begin{array}{l} j \in J \\ t \in T \end{array}$$

[total scrap available (used) for copper semi-manufacture at location j] = [scrap available (used) for tube and sheet manufacture at location j]

Capital charges for mines, smelters and refineries

$$\Phi_t^{km} = \sigma . \sum_{\substack{\tau \varepsilon T \\ \tau \leq t}} \sum_{i \varepsilon I} \sum_{m \varepsilon MM} s_i^{fm} . \left\{ \omega_{i,m}^m . S_{m,i,\tau}^m + \nu_{i,m}^m . H_{m,i,\tau}^m \right\} t \ \varepsilon \ T$$

[annualized rental charges in period t] = [scale cost of investment in all productive units m at all locations i] + [proportional cost of investments in all productive units m at all locations i]

Capital charges for wire, tubes and sheet plants

$$\Phi_t^{ks} = \sigma . \sum_{\substack{\tau \varepsilon t \\ \tau \leq t}} \sum_{j \varepsilon J} \sum_{m \varepsilon MS} s_j^{fj} . \left\{ \omega_{j,m}^s . S_{m,j,\tau}^s + \nu_{j,m}^s . H_{m,j,\tau}^s \right\} t \ \varepsilon \ T$$

[annualized rental charges in period t] = [scale cost of investment in all productive units m at all locations j] + [proportional cost of investments in all productive units m at all locations j]

Operating costs for mines, smelters and refineries

$$\Phi_t^{om} = \sum_{p \varepsilon PM} \sum_{i \varepsilon I} \zeta_i^{pi} \cdot o_i^{pm} \cdot Z_{p,i,t}^{m} \qquad t \varepsilon T$$

[operating costs for all plants i]

Operating costs for wire, tube and sheet plants

$$\Phi_t^{os} = \sum_{p \varepsilon PS} \sum_{j \varepsilon J} \zeta_j^{pj} \cdot o_{j,p}^{ps} \cdot Z_{p,j,t}^{s} \qquad t \varepsilon T$$

[operating costs for all locations j]

Transport costs

$$\Phi_t^{tr} = \sum_{c \varepsilon CIM} \sum_{i \varepsilon I} \sum_{i' \varepsilon I} \mu_{i,i',c}^{r} \cdot X_{c,i,i',t}^{i}$$

[transport cost in period t] = [transport costs of intermediate products (ore and blister)]

$$+ \sum_{c \varepsilon CFS} \sum_{j \varepsilon J} \sum_{j' \varepsilon J} \mu_{j,j'}^{fs} \cdot X_{c,j,j',t}^{fs}$$

+ [transport costs of copper semimanufactures]

$$+ \sum_{i \varepsilon J} \sum_{j \varepsilon J} \mu_{i,j}^{i} \cdot [X_{i,j,t}^{ir} + X_{i,j,t}^{fr}] \quad t \varepsilon T$$

+ [transport cost of refined copper for final use at markets and intermediate input use at semimanufacturing locations]

Tariff costs

$$\Phi_t^{tf} = \sum_{j \in J} \{ \sum_{i \in I} t_{i,j}^{r} \cdot [X_{i,j,t}^{fr} + X_{i,j,t}^{ir}]$$

[total tariffs levied on refined copper]

$$+ \sum_{j' \in J} \sum_{c \in CFS} t_{j',j}^{s} \cdot X_{c,j',j,t}^{fs} \qquad t \in T$$

+ [total tariffs levied on finished copper semi-manufactures]

Total cost

$$\Phi = \sum_{t \varepsilon T} n_t \cdot \delta_t \cdot \left\{ [\Phi_t^{km} + \Phi_t^{ks}] + \Phi_t^{om} + \Phi_t^{os} + \Phi_t^{tr} + \Phi_t^{tf} \right\}$$

[Total discounted cost] = [capital charges for mines, smelters and refineries in period t] + [capital charges for wire, tube and sheet plants in period t]

+ [operating costs for mines, smelters and refineries in period t] + [operating costs for wires, tube and sheet plants in period t]

+ [transport costs in period t] + [tariff costs in period t]

3.3 Data and Coefficients Used for the Equations

3.3.1 Market requirements

The basic data consists of consumption levels for refined copper and copper semi-manufactures (wire, tubes and rods, and sheet, plants and strip) in the aggregated market regions for the years 1990, 1995 and 2000. Section 2.7 explains the methodology used to generate these estimates. Table 2.13 presents consumption projections for 1990 and 2000.

3.3.2 Technology

Table 3.1 shows the input-output matrix with the coefficients normalized on the output. It shows the number of units of commodity c required (-) per unit of output (+) of process p. For example, to produce one unit of refined copper would require either 1.03 units of blister or scrap input. Table 3.2, the capacity utilization matrix, gives information on the number of units of capacity required on productive unit m per unit of output of process p. Capacity is expressed in units of pure copper of the higher ore grade. Thus the coefficient of 1.6 for the medium-grade ore mining process implies that this process requires 60% more capacity per ton of copper content extracted. Tables 3.3, 3.4 and 3.5, show the initial capacities and copper reserves used. The same technology is assumed to be available at all locations, except for those plants which may be able to use copper scrap.

3.3.3. Scrap availability

The amount of scrap is limited at each location. According to quality, some scrap may be used directly at

Table 3.1: INPUT-OUTPUT COEFFICIENTS

Commodities / Processes	High-Grade Ore Mining	Medium Grade Ore Mining	Smelting	Smelting Using Scrap	Refining	Refining Using Scrap	Wire Fabrication From Refined Copper	Tube & Rod Manufacture Using Refined Copper	Tube & Rod Manufacture Using Scrap	Sheet Manufacture Using Refined Copper	Sheet Manufacture Using Scrap
Ore	1	1	-1.03								
Blister			1	1	-1.03						
Refined Copper					1	1	-1.007	-.718		-.772	
Scrap for smelting				-1.03							
Scrap for refining						-1.03					
Scrap for tube manufacture									-.718		-.772
Scrap for sheet, plate and strip manufacture											
Wire							1				
Tubes and rods								1	1		
Sheets, plates and strip										1	1

Table 3.2: CAPACITY UTILIZATION MATRIX

Commodities / Processes	High-Grade Ore Mining	Medium Grade Ore Mining	Smelting	Smelting Using Scrap	Refining	Refining Using Scrap	Wire Fabrication From Refined Copper	Tube & Rod Manufacture Using Refined Copper	Tube & Rod Manufacture Using Scrap	Sheet Manufacture Using Refined Copper	Sheet Manufacture Using Scrap
Open pit	1	1.6									
Smelter		.32	1	.32							
Refinery					1	1					
Wire Unit							1				
Tube and rod unit								1	1		
Sheet, Plate and strip unit										1	1

Table 3.3: ESTIMATES OF MINE, SMELTER AND REFINERY CAPACITIES IN 1980

('000 tons)

	Open-Pit	Smelter	Refinery
Peru	590	350	230
Chile	1,330	950	810
Zambia	550	601	610
Zaire	630	430	140
Mexico-Central America	324	86	100
South Africa	490	280	150
Philippines	492	-	-
Papua New Guinea	256	-	-
Western US	1,675	1,510	480
Eastern US	576	540	1,670
Canada	1,254	610	610
Eastern Europe-USSR	1,720	1,720	1,600
Australia	260	220	210
Western Europe	608	820	1,440
Japan-Korea	112	1,180	1,220

Table 3.4: ESTIMATES OF COPPER SEMI-MANUFACTURE CAPACITIES IN 1980

('000 tons)

	Wire	Tubes and Rods	Sheets, Plates and Strip
USA	2,100	900	1,000
Mexico-Central America	105	40	15
Eastern South America	270	65	30
Western South America	55	15	9
Western Europe	1,900	1,200	900
Eastern Europe-USSR	1,200	600	350
Central America	15	5	7
South Africa	55	40	25
Other Asia	125	30	4
Japan-Korea	1,200	600	410
China	250	85	60
Australia	130	70	70
Canada	200	50	60

Table 3.5: ORE RESERVES IN 1980-81

(million tons)

	High-Grade Ore	Medium-Grade Ore
Peru	1.67	18.8
Chile	55.77	17.29
Zambia	18.01	1.21
Zaire	16.97	1.48
Mexico-Central America	3.61	20.14
South Africa	2.42	.76
Papua New Guinea		9.5
Philippines	4.93	3.97
Western US	19.38	25.87
Eastern US		5.35
Canada	8.33	9.6
Eastern Europe-USSR	38	17
Australia	6	.8
Western Europe		7.87
Japan-Korea		1.4

semimanufacturing plants, while other qualities must be first processed at smelters and refineries. This has been described in Section 2.6. Projections on scrap availability for the time horizon covered in this study were derived based on scrap potential based on past consumption coupled with historical trends.

3.3.4 Investment Costs

The investment data collected for plants was a fixed cost (million US$) to account for economies of scale size, and a variable cost (million US$ per 1,000 tons). The investment formulation conssiders that, from the maximum economies-of-scale point on to the maximum size, a plant only incurs a proportional cost. But, if the plant size is less than the economies-of-scale size, then in addition to the proportional cost, the plant must also incur a scale cost arising from the loss of economies-of-scale. Tables 2.4 and 2.6 present the investment cost estimates for copper mines and plants. The investment formulations used is as follows:

fixed cost: FC (million US$)
variable cost: VC (million US$/1,000 tons)
economies-of-scale size: $\bar{h}$
maximum size: h

Let H = expansion size
S = unused economies-of-scale
ω = scale cost
ν = proportional cost

Then,

$$\omega = \frac{FC}{\bar{h}} \qquad \begin{aligned} \nu &= \omega + VC \\ &= \frac{FC}{\bar{h}} + VC \end{aligned}$$

Expansion constraints are:

$$S + H \geq \bar{h} \cdot Y$$

$$H \leq \hat{h} \cdot Y$$

$$\Phi^{k} = \omega \cdot S + \nu \cdot H$$

where Y is the binary decision variable. If $H \geq \hat{h}$, then S is zero and the cost incurred is purely proportional, but if $H \leq \hat{h}$, then S is non-zero and an unused scale cost is also incurred in addition to the proportional cost.

Figure 3.1
The Investment Cost Function

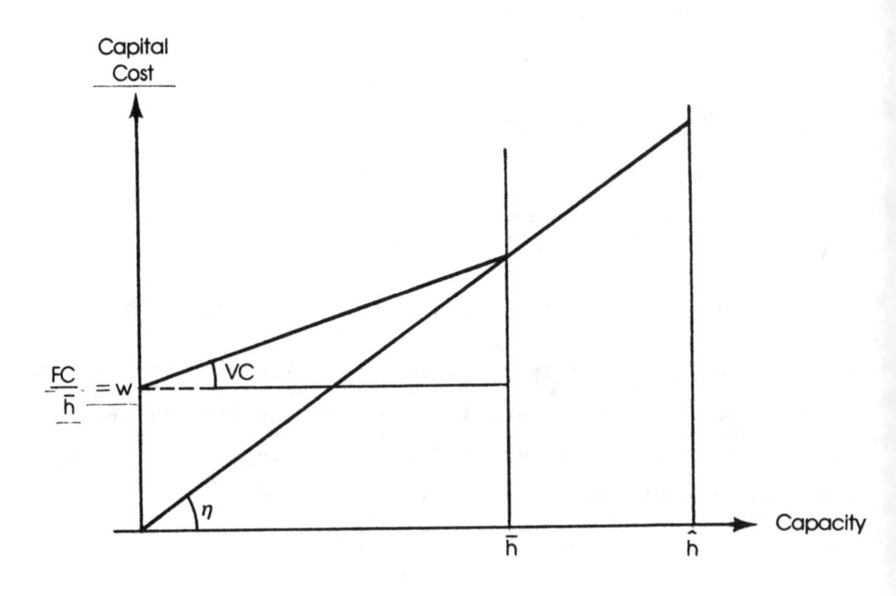

The capital recovery factor converts the investment costs of capacity construction to an even annualized stream of payments that covers the original capital charges and the interest charges on them. This stream of payments is spread out over the useful life of the productive unit.

Let, σ = capital recovery factor

ρ = discount rate

ξ = useful life of productive unit

$$\sigma = \frac{\rho(1 + \rho)^{\xi}}{(1 + \rho)^{\xi} - 1}$$

Operating Costs

Operating costs for copper processing and semi-manufactures are given in tables 2.5 and 2.7. The operating costs for semi-manufactures shown exclude the cost of refined copper. The processes using ore, blister and refined copper as inputs require one unit of capacity from the relevant productive unit for one unit of output. Therefore, the operating costs of these processes (Tables 2.5 and 2.7) indicate the cost of producing these products on a $ per ton of copper content. Processes using scrap input have additional costs, such as collection costs, and also require less capacity from the same productive unit for one unit of output (see Tables 3.1 and 3.2). The operating cost of the scrap using processes is therefore given by:

$$\begin{bmatrix}\text{units of capacity needed}\\ \text{on productive unit m per}\\ \text{unit of output}\end{bmatrix} \quad x \quad \begin{bmatrix}\text{operating costs of}\\ \text{equivalent process}\\ \text{per unit of output}\\ \text{using non-scrap}\\ \text{output}\end{bmatrix}$$

$$+ \begin{bmatrix}\text{scrap collection}\\ \text{cost}\end{bmatrix} \quad x \quad \begin{bmatrix}\text{scrap needed per}\\ \text{unit of putput}\end{bmatrix}$$

3.3.5 Tariffs

The tariffs imposed in the model assumed the rates shown in Table 2.10 and prices of US$2,000 per ton and US$2,200 per ton for refined copper and copper semi-manufactures respectively.

Table 3.6: IMPORT TARIFFS

(%)

	Refined Copper	Copper Semi-manufactures
USA	1	6
Mexico	15	15
Eastern South America	15	15
Western South America	15	15
Western Europe	0	0
Western Europe-USSR	15	15
Central Africa	15	15
South Africa	15	15
Other Asia	15	15
Japan-Korea	8.5	15
China	15	15
Australia	0	15
Canada	0	6

4. Results

4.1 Organization of the Results

This chapter presents the results and their interpretation under various assumptions. The second section of this chapter analyzes the base case, which considers the year 2000 as the target year, and includes tariffs on imports of refined copper and copper semimanufactures. In order to take into account changes in relative costs as well as errors in aggregation and cost estimation, four alternative scenarios to the base case are presented in section 4.3. Section 4.4 considers what would be the investment patterns in the copper industry if import tariffs were removed and thus this solution represents the optimal allocation of resources. Finally, section 4.5 looks at the timing of investment patterns. For this, it breaks down investments by region for 1990, 1995 and 2000.

4.2 The Base Case: Tariffs on Imports and Full Capacity Utilization on Wire Semimanufacturing Plants

The tables relating to the base case cover: a) existing capacity in 1982, investments in new capacity between 1982-2000, and total capacity in 2000 (Table 4.1), and b) interplant and final product shipments (Tables 4.2-4.7).

The results at each level of mining/processing are the results of a complex interrelationship of various factors which would be discussed accordingly. At the first stage--mining and milling--investments are closely related to the

Table 4.1: Base Case: Copper Mine and Plant Capacities and Expansions, 1982-2000

	Mines			Smelters			Refineries		
	Capacity 1982	Expansions* 1982-2000	Capacity 2000	Capacity 1982	Expansions 1982-2000	Capacity 2000	Capacity 1982	Expansions 1982-2000	Capacity 2000
Peru	400	-	400	350	-	350	230	-	230
Chile	1,330	1,458	2,788	950	157	1,107	810	265	1,075
Zambia	550	350	900	601	-	601	610	251	861
Zaire	630	292	922	430	-	430	140	-	140
Mexico-C. America	270	796	1,066	86	975	1,061	100	941	1,041
South Africa	350	-191	159	280	-	280	150	122	272
Philippines	400	-	400	-	431	431	-	418	418
Papua N. Guinea	160	315	475	-	155	155	-	150	150
Western US	1,410	852	2,262	1,510	-	1,510	480	1,092	1,572
Eastern US	360	-93	267	540	648	1,188	1,670	202	1,872
Canada	940	-44	896	610	-	610	610	-	610
USSR-E. Europe	1,720	180	1,900	1,720	925	2,645	1,600	968	2,568
Australia	260	62	322	220	-	220	210	41	251
W. Europe	380	-	380	820	371	1,191	1,440	459	1,899
Japan-Korea	70	-	70	1,180	-	1,180	1,220	669	1,889
TOTAL	9,230	3,977	13,207	9,297	3,662	12,959	9,270	5,578	14,848

*negative expansions and retirements.

Table 4.1 (Cont.): Base Case: Copper Mine and Plant Capacities and Expansions, 1982-2000

	Wire Plants			Sheet Plants			Tube Plants		
	Capacity 1982	Expansions 1982-2000	Capacity 2000	Capacity 1982	Expansions 1982-2000	Capacity 2000	Capacity 1982	Expansions 1982-2000	Capacity 2000
Mexico-C. America	105	800	905	15	60	75	40	47	87
South Africa	55	129	184	25	19	43	40	25	65
Canada	200	-	200	60	-	60	50	41	91
USSR-E. Europe	1,200	-	1,200	350	405	755	600	344	944
Australia	130	-	130	70	-17	53	70	-9	61
W. Europe	1,900	-	1,900	900	-172	728	1,200	490	1,690
Japan-Korea	1,200	-	1,200	410	304	714	600	297	897
US	2,100	-	2,100	1,000	-180	820	900	120	1,020
Eastern S. America	270	155	425	30	145	175	65	142	207
Western S. America	55	204	259	9	30	39	15	29	44
Central Africa	15	800	815	7	4	11	5	10	15
Other Asia	125	71	196	4	72	76	30	68	98
China	250	660	910	60	74	134	85	83	168
TOTAL	7,605	2,819	10,424	2,940	744	3,684	3,700	1,687	5,387

Table 4.2: BASE CASE: SHIPMENTS OF COPPER CONCENTRATES, 2000

('000 tons)

	Peru	Chile	Zambia	Zaire	Mexico-C. America	South Africa	Philippines	Papua N. Guinea	Western US	Eastern US	Canada	USSR-E. Europe	Australia	Western Europe	Japan-Korea	TOTAL
Peru	360									40						400
Chile		1,140			27					908		713				2,788
Zambia			619	281												900
Zaire				162		129						111		520		922
Mexico-S. America					1,066											1,066
South Africa						159										159
Philippines							400									400
Papua N. Guinea								159							316	475
Western US									1,521						741	2,262
Eastern US										267						267
Canada											628			268		896
USSR-E. Europe												1,900				1,900
Australia							44						224		54	322
W. Europe														380		380
Japan-Korea															70	70
TOTAL	360	1,140	619	443	1,093	288	444	159	1,521	1,215	628	2,724	224	1,168	1,181	13,207

Table 4.3: BASE CASE: SHIPMENTS OF BLISTER COPPER, 2000

('000 tons)

	Peru	Chile	Zambia	Zaire	Mexico-C. America	South Africa	Philippines	Papua N. Guinea	Western US	Eastern US	Canada	USSR-E. Europe	Australia	Western Europe	Japan-Korea	TOTAL
Peru	237									113						350
Chile		1,107														1,107
Zambia			601													601
Zaire			286	144												430
Mexico-C. America					1,061											1,061
S. Africa						280										280
Philippines							431									431
Papua New Guinea								155								155
Western US									1,126						384	1,510
Eastern US										1,188						1,188
Canada										69	541					610
USSR-E. Europe												2,645				2,645
Australia													220			220
W. Europe														1,191		1,191
Japan-Korea															1,180	1,180
TOTAL	237	1,107	887	144	1,061	280	431	155	1,126	1,370	541	2,645	220	1,191	1,564	12,959

Table 4.4: BASE CASE: SHIPMENTS OF REFINED COPPER, 2000

('000 tons)

	Mexico-C. America	South Africa	Canada	USSR-E. Europe	Australia	Western Europe	Japan-Korea	USA	E.S. America	W.S. America	Central Africa	Other Asia	China	TOTAL
Peru						230								230
Chile									747	328				1,075
Zambia						19					842			861
Zaire						140								140
Mexico-C. America	1,041													1,041
S. Africa		272												272
Philippines													418	418
Papua New Guinea													150	150
Western US								716				265	591	1,572
Eastern US								1,872						1,872
Canada			303			307								610
USSR-E. Europe				2,568										2,568
Australia					176							75		251
W. Europe						1,899								1,899
Japan-Korea							1,889							1,889
TOTAL	1,041	272	303	2,568	176	2,595	1,889	2,588	747	328	842	340	1,159	14,848

Table 4.5: BASE CASE: SHIPMENTS OF COPPER WIRE, 2000

('000 tons)

	Mexico-C. America	South Africa	Canada	USSR-E. Europe	Australia	Western Europe	Japan-Korea	USA	E.S. America	W.S. America	Central Africa	Other Asia	China	TOTAL
Mexico-C. America	174			400		331								905
South Africa		120		63										183
Canada			190											190
USSR-E. Europe				1,200		10								1,210
Australia					129									129
Western Europe						1,900								1,900
Japan-Korea							1,200							1,200
USA				52				2,048						2,100
Eastern S. America									425					425
Western S. America				172						87				259
C. Africa						782					33			815
Other Asia												196		196
China							583						327	910
TOTAL	174	120	190	1,887	129	3,023	1,783	2,048	425	87	33	196	327	10,422

Table 4.6: BASE CASE: SHIPMENTS OF COPPER SHEET, PLATE AND STRIPS, 2000

('000 tons)

	Mexico-C. America	South Africa	Canada	USSR-E. Europe	Australia	Western Europe	Japan-Korea	USA	E.S. America	W.S. America	Central Africa	Other Asia	China	TOTAL
Mexico-C. America	65		10											75
South Africa		44												44
Canada			60											60
USSR-E. Europe				755										755
Australia					53									53
Western Europe						728								728
Japan-Korea							713							713
USA								820						820
Eastern S. America									174					174
Western S. America			6							33				39
Central Africa											11			11
Other Asia												76		76
China													134	134
TOTAL	65	44	76	755	53	728	713	820	174	33	11	76	134	3,682

Table 4.7: BASE CASE: SHIPMENTS OF COPPER TUBES AND RODS, 2000

('000 tons)

	Mexico-C. America	South Africa	Canada	USSR-E. Europe	Australia	Western Europe	Japan-Korea	USA	E.S. America	W.S. America	Central Africa	Other Asia	China	TOTAL
Mexico-C. America	87													87
South Africa		65												65
Canada			91											91
USSR-E. Europe				944										944
Australia					61									61
Western Europe						1,690								1,690
Japan-Korea							897							897
USA								1,020						1,020
Eastern S. America									207					207
Western S. America										44				44
Central Africa											15			15
Other Asia												98		98
China													168	168
TOTAL	87	65	91	944	61	1,690	897	1,020	207	44	15	98	168	5,387

availability of low cost/high grade copper reserves. Transport costs take a secondary role, exerting an influence only when the choice is among comparable copper deposits. The availability of large high quality copper deposits result in high investment levels in Chile and Zambia. For the former, mining capacity could more than double, increasing from 1.3 million tons per year in 1982 to about 2.8 million tons per year in 2000. For the same period, mine capacity in Zambia could rise from 550,000 tpy to 900,000 tpy. The influence of by-products is decisive in determining new investments in Papua New Guinea and Zaire. Thus in Papua New Guinea, where gold is a major by-product, mining capacity expands from 160,000 tpy in 1982 to 475,000 tpy by 2000. Zaire, where cobalt contributes substantially to revenues at copper mines, expands its mine capacity from 630,000 tpy in 1982 to about 920,000 tpy in 2000. Mining investments in the U.S. are a somewhat more marginal case since new mines have a less important cost advantage over potential mines in other areas. Mining capacity in the U.S. would increase from 1.7 million tpy in 1982 to about 2.5 million tpy in 2000. All the ore mined in the US is locally smelted. The Soviet Bloc shows a less significant increase in capacity--from 1.7 million tpy in 1982 to 1.9 million tpy in 2000--to supply its own smelters.

According to the above investment patterns, by 2000, the relative ranking of copper producing countries may change substantially an outcome which is not unexpected. Chile could easily surpass the U.S. and the Soviet bloc in copper mining output, while the latter would become second and third in importance as copper producers. Mexico and Central America could become a very important copper producing area, and Zaire, Zambia and Canada would maintain their role of important copper producers. These seven regions/countries together would account for 11 million tons of mining output by 2000, or about 85% of world production. It should be noted, however, that the above results could change if new copper reserves are discovered. Besides, as noted earlier, some investment decisions may be marginal. Investments in Mexico-Central America and the U.S. could be lower, counterbalanced by expansion in Peru, the Soviet bloc and additional expansions in Chile, all of which have abundant reserves of lower quality ore. This would not affect significantly the costs of productions. The total costs of copper smelting are only slightly lower in industrial countries with respect to those at developing countries. Lower investment costs at industrial countries, due to less infrastructure requirements, are almost compensated with higher operating costs due mainly to higher wages. Thus, investment patterns in copper smelting are basically determined by locational factors. Less Developed Countries build new smelters to process their own copper ores and to supply their local refineries. Thus, Chile expands smelter capacity by 157,000 tpy reaching a total capacity of 1.1 million tpy. Mexico-Central America expand capacity by 975,000 tpy to 1.06 million tpy. Philippines and Papua New Guinea became producers of blister copper by building up 431,000 tpy and 155,000 tpy of smelter capacity respectively. All these countries ship all their output to their own refineries.

Although the industrial countries build new smelters to keep supplying their refineries, they use both local and imported copper conentrates. The Eastern US increases its smelting capacity by 648,000 tpy to supply its own refineries, reaching 1.19 million tpy capacity, using as feed 267,000 tpy of its own concentrates and 908,000 tpy of concentrates from Chile and 40,000 tpy from Peru. The Soviet bloc increases its smelting capacity substantially by about 925,000 tpy reaching 2.6 million tpy capacity, to match its new refining capacity, using copper concentrates from: own mines (1.9 million tpy), Chile (712,000 tpy) and Zaire (111,000 tpy). Western Europe expands its copper smelting capacity by 371,000 tpy to reach a total capacity of 1.2 million tpy, using local concentrates (380,000 tpy), as well as imports from Zaire (520,000 tpy) and Canada (268,000 tpy).

Investments in copper refineries are determined by transport costs and import tariffs. The latter are not applied on refined copper exported to Western Europe, Australia and Canada. The main result are as follows: Chile increases refinery capacity by 265,000 tpy, reaching 1.07 million tpy to supply semimanufacturing plants in Western and Eastern South America. Zambia builds 251,000 tpy of copper refining capacity, reaching 861,000 tpy, mainly to supply Central Africa. Mexico-Central America add 941,000 tpy to reach a toal refining capacity of 1.04 million tpy to supply their own semimanufacturing plants. South Africa expands refining capacity by 122,000 tpy reaching 272,000 tpy in order to supply its semimanufacturing plants. The Philippines and Papua New Guinea become producers of refined copper by building copper refineries of 418,000 tpy and 150,000 tpy capacity respectively, supplying China.

Industrial countries expand refinery capacity and use both local and imported copper blister as feedstock. The Western US expands refinery capacity by 1.09 million tpy reaching 1.5 million tpy total, to match its smelter capacity and supplies semimanufacturing plants in the U.S., China and Other Asia. The Eastern U.S. expands refinery capacity by 202,000 tpy, reaching 1.9 million tpy total, using as inputs local blister plus 113,000 tpy imported from Peru and 70,000 tpy from Canada and supplies its own semimanufacturing plants. The Soviet bloc matches its smelter and semimanufacturing capacity by expanding its refineries by 968,000 tpy reaching 2.6 million tpy.

Australia uses its own blister copper, expands refinery capacity by 41,000 tpy to 251,000 tpy, and exports 62,000 tpy of refined copper to Other Asia. Western Europe matches its smelter capacity by adding 459,000 tpy of refinery capacity, for a total of 1.9 million tpy to supply its own semimanufacturing plants. Japan-Korea import 383,000 tpy blister from Western U.S., to add to 1.9 million tpy of their own blister, expanding refinery capacity by 669,000 tpy to 1.9 million tpy to supply their own semimanufacturing plants.

Because the sheet/plates and tube semimanufacturing plants may process copper scrap, the availability of copper scrap (at prices lower than that of refined copper) is a major factor in determining new investments, particularly in industrial countries where it exists in abundance. The trade off between low operating costs/high capital costs for developing countries and high operating costs (higher wages)/low capital costs for industrial countries is the decisive factor for plants which are to use primary refined copper. This trade-off takes place mainly in the supply of semimanufactures to industrial countries, because semimanufacturing plants in developing countries aimed at supplying local markets are protected both by tariffs and low labor costs.

New wire plants are located only in less developed countries as they are the most labor intensive of the semimanufacturing plants (Table 4.1). Investments occur in those developing countries which hold a strategic position with respect to markets as well as refined copper supplies. Thus, Mexico-Central America expand by 800,000 tpy of wire plant capacity reaching a total of 905,000 tpy, and supplying the Soviet bloc (687,00 tpy) and Western Europe (182,000 tpy). Central Africa expands from almost nil to 815,000 tpy to supply Western Europe. China expands by 660,000 tpy reaching 910,000 tpy capacity to supply Japan-Korea (582,000 tpy) and its own market. Eastern South America expands by 155,000 tpy for a total of 425,000 tpy for its own market, while Western South America increases capacity by 204,000 tpy reaching 259,000 tpy to supply its own market (87,000 tpy) and USSR-E. Europe (171,000 tpy). South Africa expands by 129,000 tpy to arrive at 184,000 tpy total capacity supplying its own market (119,000 tpy) and USSR-E. Europe (63,000). The results for new sheet/strip and tube semimanufacturing plants are more straightforward. Both the availability of scrap and tariff protection determine that new capacity is to be established in such a way that local markets are satisfied. In the case of sheet/strip plants, capacity underutilization arises as industrial countries are not able to export competitively. Thus the U.S. uses 82% of sheet plant capacity, Japan-Korea's capacity utilization is 81%, and Australia's decreases to 76%.

4.3 Four Alternative Scenarios

This section covers four alternative scenarios to the base case. These are:

Variant a) Free Capacity Utilization at Wire Plants;
Variant b) 25% cost reduction at mines in Chile, Peru and USSR-E. Europe, to account for errors in aggregation and in cost estimation;
Variant c) Increases in investment costs in developing countries to account for possible higher infrastructure costs as well as risk and other factors;
Variant d) Increases in operating costs at LDCs to account for possible lower productivity, future cost escalation in real wages, and in general to test the "worse" possible scenario for developing countries.

Table 4.8 presents investment patterns for the base case and the four scenarios mentioned above. The latter four will be discussed in turn, and will be compared to the base case.

The first alternative scenario eliminates the restriction that all existing wire semimanufacturing plants are to operate at full capacity. With this restriction removed, wire plants in the US operate at less than 40% capacity, compensated for by investments in Western South America and South Africa. Thus there are even larger investments in new smelting and refining capacity in Chile and South Africa than there were in the base case. The Eastern US, Papua New Guinea, Western Europe and Zambia, compensate for the above increases by showing lower or no investments in smelters and refineries. This scenario is unlikely as the volume of capacity closures and openings is very large and would probably be resisted by integrated copper companies and would also be rendered difficult by the amount of investment funds required for new plants. However, an intermediate solution between the base case and variant a is more likely.

Table 4.8: BASE CASE AND VARIANTS: COPPER MINE AND PLANT CAPACITY EXPANSION - 2000

		Variant A	Variant B	Variant C	Variant D
	Base Case	Base Case Free Cap. Ut.	Cost Reduction Chile Peru-USSR	25% Increase Invest. Cost LDCs	25% Increase Operating Costs LDCs
Mines					
Peru	-	-	624	-	-
Chile	1,458	1,458	1,458	1,458	1,458
Zambia	350	350	350	350	350
Zaire	292	292	292	238	238
Mexico-C. America	796	796	-	-	-
S. Africa	-191	-191	-191	-191	-191
Philippines	-	-	-	-	-
Papau N. Guinea	315	315	315	315	315
Western US	852	852	196	852	852
Eastern US	-93	-93	-93	-93	-93
Canada	-44	-44	-44	-44	-44
USSR-E. Europe	180	180	1,030	1,030	1,030
Australia	62	62	40	62	62
W. Europe	-	-	-	-	-
Japan-Korea	-	-	-	-	-
TOTAL	3,977	3,977	3,977	3,977	3,977
Smelters					
Peru	-	-	-	-	-
Chile	157	776	157	-	-
Zambia	-	-	-	-	-
Zaire	-	-	-	-	-
Mexico-C. America	975	949	975	-	-
S. Africa	-	593	-	-	-
Philippines	431	383	-	-	-
Papau N. Guinea	155	-	-	-	-
Western US	-	-	-	-	-
Eastern US	648	-	1,234	1,576	1,236
Canada	-	-	-	-	-
USSR-E. Europe	925	961	925	950	1,638
Australia	-	-	-	150	150
W. Europe	371	-	371	551	638
Japan-Korea	-	-	-	435	-
TOTAL	3,662	3,662	3,662	3,662	3,662

Table 4.8 (Cont.) BASE CASE AND VARIANTS: COPPER MINE AND PLANT CAPACITY EXPANSION - 1982 - 2000

	Base Case	Variant A Base Case Free Cap. Ut.	Variant B Cost Reduction Chile Peru-USSR	Variant C 25% Increase Invest. Cost LDCs	Variant D 25% Increase Operating Costs LDCs
Refineries					
Peru	-	-	-	-	-
Chile	265	866	265	112	-
Zambia	251	150	251	104	-
Zaire	-	-	-	-	-
Mexico-C. America	941	941	941	820	159
S. Africa	122	798	122	122	58
Philippines	418	372	-	-	-
Papua N. Guinea	150	-	-	-	-
Western US	1,092	572	1,092	212	447
Eastern US	202	-	770	926	720
Canada	-	-	-	68	-
USSR-E. Europe	968	1,003	968	992	1,660
Australia	41	41	41	187	187
W. Europe	459	166	459	780	1,033
Japan-Korea	669	669	669	1,255	1,314
TOTAL	5,578	5,578	5,578	5,578	5,578
Wire Plants					
Mexico-C. America	800	800	800	688	70
S. Africa	129	800	129	129	65
Canada	-	-10	-	-	-
USSR-E. Europe	-	35	-	24	687
Australia	-	-	-	-	-
W. Europe	-	-	-	-	800
Japan-Korea	-	-	-	583	584
USA	-	-1,293	-	-	-
E.S. America	155	155	155	155	155
W.S. America	204	800	204	292	32
C. Africa	800	800	800	800	278
Other Asia	71	71	71	71	71
China	660	661	660	77	77
TOTAL	2,819	2,819	2,819	2,819	2,819

Table 4.8 (Cont.) BASE CASE AND VARIANTS: COPPER MINE AND PLANT CAPACITY EXPANSION - 1982-2000

	Base Case	Variant A Base Case Free Cap. Ut.	Variant B Cost Reduction Chile Peru-USSR	Variant C 25% Increase Invest. Cost LDCs	Variant D 25% Increase Operating Costs LDCs
Sheet Plants					
Mexico-C. America	60	60	60	50	-
S. Africa	19	19	19	19	19
Canada	-	-	-	-	-
USSR-E. Europe	405	405	405	405	405
Australia	-17	-17	-17	-17	-
W. Europe	-172	-172	-172	-172	-51
Japan-Korea	304	304	304	304	377
USA	-180	-180	-180	-170	-
E.S. America	145	145	145	145	-30
W.S. America	30	30	30	30	24
C. Africa	4	4	4	4	-
Other Asia	72	72	72	72	-
China	74	74	74	74	-
TOTAL	744	744	744	744	744
Tube Plants					
Mexico-C. America	47	47	47	47	47
S. Africa	25	25	25	25	25
Canada	41		41	41	4141
USSR-E. Europe	344	344	344	344	344
Australia	-9	-9	-9	-9	-
W. Europe	490	490	490	490	490
Japan-Korea	297	297	297	297	297
USA	120	120	120	120	120
E.S. America	142	142	142	142	142
W.S. America	29	29	29	29	29
C. Africa	10	10	10	10	10
Other Asia	68	68	68	68	59
China	83	83	83	83	83
TOTAL	1,687	1,687	1,687	1,687	1,687

Next, the effect of lower mining costs in Chile, Peru and the Soviet Union-Eastern Europe is considered. Under this scenario, Peru expands mine capacity by 624,000 tpy in order to exploit lower ore grades, while the Soviet Union-E. Europe expand by 1.03 million tpy instead of the previous expansion of 180,000 tpy. To compensate for these expansions, Mexico-Central America show no mining investments (previously they showed an expansion of 796,000 tpy), while the Western United States expands mines by only 196,000 tpy (as opposed to the previous expansion of 852,000 tpy). Since with these expansion patterns the Soviet Union-E. Europe become self-sufficient in copper, Chile's exports of ore are diverted to the Eastern United States. As a consequence, smelters in the Eastern United States expand by 1.2 million tpy versus the previous 648,000 tpy, while refineries in the same areas expand by 770,000 tpy instead of 202,000 tpy of the base case. Besides, there are no investments in smelters and refineries in the Philippines and Papua New Guinea, as they replace Western US in supplying copper concentrates to Japan-Korea. Western-US refineries remain at 1.09 million tpy but shipments are destined to China and Other Asia instead of the local market which was in the base case/situation. Semimanufacturing plants are not affected by any of these changes.

An increase of 25% on investment costs at developing countries results, as expected, in major displacements in investments from developing countries to industrial countries. At the mining stage, the Soviet Union-Eastern Europe expands considerably more than before, by 1.03 million tpy (vs. base case: 180,000 tpy), while Mexico-Central America do not show investments in copper mines under this scenario, and Zaire shows a lower investment than in the base case (238,000 tpy instead of the original 292,000 tpy). Investments in new smelter capacity take place exclusively in industrial countries, while at the refining stage, investments at developing countries are lower than in the base case but are still significant. Chile and Zambia show investments in new refineries equal to about one-half the volumes shown in the base case. Mexico/Central America show a somewhat lower but still significant level of investment in refineries (820,000 tpy); but the Philippines and Papua New Guinea--in contrast to the base case--do not show any capacity expansions in copper refineries. At the semimanufacturing stage, only wire plants show significant differences with the base case: Mexico-Central America show a capacity expansion of 688,000 tpy (vs. 800,000 tpy in the base case), but Western South America increases wire plant capacity by more than in the base case (292,000 tpy vs. 204,000 tpy) as this region gets an increased supply of refined copper from Chile. A trade-off arises between China and Japan-Korea: China expands wire semifabricating plants by 77,000 tpy instead of 660,000 tpy of the base case, while Japan/Korea covers its local market by expanding wire plant capacity by 583,000 tpy vs. no expansion in the base case.

An increase in operating costs at developing countries by 25% has a larger effect than the previous scenario with higher capital costs. At the mining stage the situation is exactly the same as that of the previous scenario, with the USSR-E. Europe showing a higher investment level than in the base case at the expense of Mexico-C. America and Zaire. New smelters are located only in industrial countries but geographical investment patterns differ from those of the scenario with higher capital costs. Thus, the Eastern US requires 1.2 million tpy additional smelter capacity vs. 1.6 million tpy of the higher capital costs case because the difference is supplied to Eastern US refineries by the Western US, since blister shipments from the latter to Mexico-C. America are lower than in the increased capital cost scenario. Smelter capacity expansion in USSR-E. Europe is 1.6 million tpy vs. 950,000 tpy in the previous scenario. Western Europe expands smelters by 638,000 tpy vs. 551,000 tpy in the increased caital cost case. Japan-Korea do not show capacity expansions in smelters, because their refineries use blister from the Western US, which in the previous scenario was shipped to Mexico-Central America refineries. Investments in copper refineries at developing countries are substantially reduced but not eliminated in this scenario. Mexico-C. America and South Africa show capacity expansions at refineries of 159,000 tpy and 58,000 tpy respectively in order to supply local semimanufacturing plants. Chile and Zambia show no capacity expansions at refineries because their semimanufacturing plant capacity is much lower than in the higher capital cost scenario. The USSR/E. Europe and Western Europe show refinery expansions significantly higher than in the increased capital cost scenarios, with 1.66 million tpy and 1.03 million tpy respectively to match higher copper capacity. With respect to wire semimanufacturing plants, expansions at developing countries under this scenario are limited to what is required for their own markets. The only exception is Central Africa which supplies in part the Western European market. Among industrial countries, the main difference with respect to the scenario with higher capital costs is for the USSR/E. Europe which expands by 687,000 tpy. For copper sheet semimanufacturing plants, this scenario shows low investment levels at developing countries--only South Africa and Western South America expand capacity by 19,000 tpy and 24,000 tpy respectively. In contrast with the other scenarios, the United States operates at full capacity and exports copper sheets to Mexico, Canada and Eastern South America. In general, developing countries import substantial amounts of copper sheets from industrial countries. The results under this scenario for copper tube plants are practically the same as for the other cases. Almost all production of copper tubes is shipped to local markets.

4.4 The Case of No Tariffs on Refined Copper and Copper Semimanufactures

Table 4.9 shows the investment patterns which would result by the year 2000 in the absence of tariffs, and contrasts these results with the base case with tariffs. Mining investments are not affected as differences in production costs are substantial at this stage. On the other hand, investments on smelters are affected by the elimination of tariffs on copper although tariffs are not applied directly on blister copper. Thus, Chile shows no smelter capacity expansion, while refinery capacity expansion is reduced in that country to 112,000 tpy. The main reason is that investments on wire semimanufacturing plants at Western South America decrease from 204,000 tpy to 32,000 tpy as they lose the USSR-E. Europe market to Mexico- Central America. By contrast, Peru shows refinery capacity expansion of 110,000 tpy to supply the United States vs. no expansion in the base case. Investments in wire plants at Eastern South America are somewhat higher than in the base case in order to supply the USSR-E. Europe.

Table 4.9 COMPARISON BETWEEN VERSIONS WITH AND WITHOUT TARIFFS INVESTMENT IN NEW CAPACITY 1982-2000

('000 metric tons)

	Mines		Smelters		Refineries		Wire Plants		Sheet Plants		Tube Plants	
	Tariffs	No Tariffs	Tariffs	No Tariffs	Tariffs	No Tariffs	Tariffs	No Tariffs	Tariffs	No Tariffs	Tariffs	No Tariffs
Peru	-	-	-	-	-	110	-	-	-	-	-	-
Chile	1,458	1,458	157	-	265	112	-	-	-	-	-	-
Western S. America	-	-	-	-	-	-	204	32	30	30	29	30
Eastern S. America	-	-	-	-	-	-	155	174	145	145	142	142
Mexico-C. America	796	796	975	1,022	941	987	800	354	60	551	47	209
Western US	852	852	-	-	1,092	1,465	-	-	-	-	-	-
Eastern US	-93	-93	648	554	202	-	-	-	-	-	-	-
US	-	-	-	-	-	-	-	-	-180	-604	120	-
Canada	-44	-44	-	-	-	-	-	-	-	-60	41	-
Zambia	350	350	-	-	251	846	-	-	-	-	-	-
Zaire	292	292	-	613	-	-	-	-	-	-	-	-
C. Africa	-	-	-	-	-	-	800	659	4	391	10	842
S. Africa	-191	-191	-	-	122	122	129	-	19	30	25	25
Philippines	-	-	431	470	418	456	-	-	-	-	-	-
Papua N. Guinea	315	315	155	155	150	150	-	-	-	-	-	-
Japan-Korea	-	-	-	-	669	296	-	-	304	-183	297	-
China	-	-	-	-	-	-	660	800	74	608	83	380
Other Asia	-	-	-	-	-	-	71	800	72	72	68	68
Australia	62	62	-	-	41	41	-	-	-17	-64	-9	-9
USSR-E. Europe	180	180	925	348	968	408	-	-	405	-	344	-
Western Europe	-	-	371	500	459	585	-	-	-172	-172	490	-
	3,977	3,977	3,662	3,662	5,578	5,578	2,819	2,819	744	744	1,687	1,687

Mexico-C. America is a case where the removal of tariffs benefits one activity at the expense of another one. Thus, there is about 5% higher investment in smelter and refineries since although wire semimanufacturing plants expand capacity by only 354,000 tpy in contrast to 800,000 tpy in the base case, sheet semimanufacturing plants increase capacity by 551,000 tpy versus 60,000 tpy in the base case, and tube semimanufacturing plants expand by 209,000 tpy vs. 47,000 tpy in the base case. In the case of wire semimanufactures, Mexico-C. America lose the W. Europe market and a third of the USSR-E. Europe market, but this loss is more than recovered by the gain of US and Canadian markets for sheet and tube copper semimanufactures.

The US is negatively affected by the removal of tariffs. The only higher level of investment with respect to the case with tariffs is at Western US which increases refinery capacity by 1.46 million tons instead of 1.1 million tons of the base case, in order to ship a higher volume of refined copper to semifabricating plants in China and Other Asia. Mexico-C. America become under this scenario, the main suppliers of copper sheets and tubes to the US at the expense of local semifabricating plants. The same situation applies to Canada which shows a reduced production level of copper sheets and tubes as it imports these products from Mexico/Central America.

Zambia shows a higher level of investment in copper refineries--846,000 tpy vs. 251,000 tpy in the base case--in order to supply large expansions in sheet and tube semimanfacturing plants in Central Africa (391,000 tpy and 842,000 tpy respectively), while in the base case investments in these were negligible. Investments at wire plants for Central Africa are somewhat lower than in the base case, because semifabricating plants in China and Other Asia obtain low cost supplies of refined copper from Western US, and therefore expand at a higher level than in the base case, and partially replace Central Africa at Western European markets. Sheet and tube products from Central Africa gain access to USSR-E. Europe markets. Shipments of wire semimanufactures from Other Asia also reach South Africa which shows no capacity expansion in wire plants, in contrast to the case with tariffs in which it expanded by 129,000 tpy.

China and Other Asia benefit substantially by the removal of tariffs. China obtains additional supplies of refined copper from Western U.S. and the Philippines, while Other Asia is supplied by Western U.S., which in the case with tariffs supplied a larger volume to Japan-Korea. Thus, China expands wire plant capacity by 800,000 tpy (vs. 660,000 tpy in the base case), sheet plant capacity by 608,000 tpy (74,000 tpy in the base case), and tube plant capacity by 380,000 tpy (83,000 tpy in the base case). Other Asia benefits exclusively in the case of wire plant capacity, expanding by 800,000 tpy instead of 71,000 tpy which resulted in the base case. As mentioned above, the availability of additional refined copper from Western US permits Other Asia and China to take away Western European copper wire markets from Central Africa. The removal of tariffs gives China access to Japan-Korea and Australian tube and sheet markets.

The USSR-E. Europe, when compared to the case with tariffs, shows lower investment levels for copper processing and no investment in copper semifabricating plants as their market is supplied by Central Africa.

Western Europe obtains a higher supply of copper concentrates than in the base case, expanding copper smelters by 500,000 tpy vs. 371,000 tpy originally and copper refineries by 585,000 tpy vs. 459,000 tpy in the base case. Thus, they reduce their imports of refined copper for their semifabricating plants. By contrast, Western Europe shows no capacity expansion in tube plants, importing copper tubes from Central Africa while in the base case it relied totally on domestic production.

In conclusion, it may be said that the removal of tariffs benefits on average the developing countries as their labor cost component is lower than for industrial countries. Thus the former gain access to the large markets of industrial countries. On the other hand, not all developing countries benefit from this action because the limited supply of low cost copper concentrates are subject to redistribution and thus copper availability is reduced at some semifabricating plants (i.e., wire plants at Western South America, Mexico-Central America and Central Africa), restricting the potential for capacity expansion.

4.5 The Dynamic Case: 1990, 1995 and 2000

Apart from the patterns of investment from the regional distribution point of view, it is interesting to consider the timing of these investments. The present scenario analyzes this dimension. Table 4.10 presents the investment patterns associated with the optimum solution, while Table 4.11 shows the corresponding production levels.

Mining investments occur early- by 1990-at Chile, Zaire, Papua New Guinea and USSR-E. Europe. Abundance of low cost/high quality copper reserves permit Chile to continue expanding in 1995, becoming the most important world

Table 4.10: INVESTMENT PATTERNS 1990, 1995 AND 2000

	1990						1995					
	Mines	Smelters	Refineries	Wire Plants	Sheet Plants	Tube Plants	Mines	Smelters	Refineries	Wire Plants	Sheet Plants	Tube Plants
Peru	-	-	-	-	-	-	-	-	-	-	-	-
Chile	366	-	-	-	-	-	1,092	-	114	-	-	-
Western So. America	-	-	-	20	24	21	-	-	-	177	-	-
Eastern So. America	-	-	-	17	88	75	-	-	-	62	26	30
Mexico-Central America	-	210	110	20	29	22	-	438	600	416	30	34
Western US	-	-	679	-	-	-	40	-	302	-	-	-
Eastern US	-93	-	-	-	-	-	-	668	-	-	-	-
US	-	-	-	-	-346	-79	-	-	-	-	82	79
Canada	-44	-	-	-	-	22	-	-	-	-	-	-
Zambia	-	-	-	-	-	-	350	-	251	-	-	-
Zaire	218	-	-	-	-	-	-	-	-	-	-	-
C. Africa	-	-	-	229	-	-	-	-	-	330	-	-
S. Africa	-191	-	122	34	-	-	-	-	-	102	-	21
Philippines	-	-	-	-	-	-	-	-	-	-	-	-
Papua N. Guinea	315	-	-	-	-	-	-	-	-	-	-	-
Japan-Korea	-	-	409	-	86	24	-	-	259	-	99	124
China	-	-	-	8	46	47	-	-	-	33	-	17
Other Asia	-	-	-	-	48	36	-	-	-	36	11	14
Australia	40	-	-	-	-28	-	-	-	-	-	12	6
USSR-E. Europe	180	325	620	-	212	102	-	198	161	-	90	112
Western Europe	-	-	-	-	-314	159	-	150	-	-	67	156
TOTAL	791	535	1,940	328	-155	407	1,482	1,454	1,687	1,156	417	593

Table 4.10 (Cont.): INVESTMENT PATTERNS 1990, 1995 AND 2000

	2000					
	Mines	Smelters	Refineries	Wire Plants	Sheet Plants	Tube Plants
Peru	-	-	-	-	-	-
Chile	-	150	144	-	-	-
Western So. America	-	-	-	-	-	8
Eastern So. America	-	-	-	76	31	37
Mexico-Central America	818	808	697	800	30	-
Western US	812	-	258	-	-	-
Eastern US	-	-	221	-	-	-
US	-	-	-	-	85	112
Canada	-	-	-	-	-	19
Zambia	-	-	-	-	-	-
Zaire	74	-	-	-	-	-
C. Africa	-	-	-	-	-	-
S. Africa	-	-	-	-	-	14
Philippines	-	150	-	-	-	-
Papua N. Guinea	-	414	402	-	-	-
Japan-Korea	-	-	-	-	118	149
China	-	-	-	424	28	19
Other Asia	-	-	-	35	14	17
Australia	-	-	41	-	-2	7
USSR-E. Europe	-	150	187	-	104	130
Western Europe	-	-	-	-	75	174
TOTAL	1,704	1,672	1,950	1,335	483	686

Table 4.11: PRODUCTION PATTERNS 1990, 1995 AND 2000

	Mines			Smelters			Refineries		
	1990	1995	2000	1990	1995	2000	1990	1995	2000
Peru	400	400	400	350	350	350	230	230	230
Chile	1,696	2,789	2,789	950	950	1,100	810	924	1,068
Zambia	550	900	900	601	601	601	610	861	861
Zaire	848	848	922	430	430	430	140	140	140
Mexico-C. America	270	270	1,088	295	734	1,542	210	810	1,508
South Africa	159	159	159	280	280	280	272	272	272
Philippines	400	400	400	-	-	150	-	-	-
Papua New Guinea	475	475	475	-	-	414	-	-	402
Western US	1,410	1,450	2,262	1,557	1,567	1,580	1,159	1,461	1,718
Eastern US	268	268	268	552	1,223	1,226	1,670	1,670	1,891
Canada	897	897	897	610	610	610	610	610	610
USSR-E. Europe	1,900	1,900	1,900	2,045	2,243	2,393	2,220	2,381	2,568
Australia	300	300	300	224	224	224	210	210	252
W. Europe	380	380	380	921	1,081	1,090	1,440	1,440	1,440
Japan-Korea	70	70	70	1,229	1,239	1,252	1,629	1,888	1,888
TOTAL	10,023	11,506	13,210	10,044	11,532	13,242	11,210	12,897	14,848

Table 4.11 (Cont.): PRODUCTION PATTERNS 1990, 1995 AND 2000

	Wire Plants			Sheet Plants			Tube Plants		
	1990	1995	2000	1990	1995	2000	1990	1995	2000
Mexico-C. America	125	541	1,341	44	74	104	62	96	96
South Africa	89	190	190	25	25	25	40	61	76
Canada	200	200	200	60	60	60	72	72	91
USSR-E. Europe	1,200	1,200	1,200	562	651	755	702	814	943
Australia	130	130	130	42	55	53	48	54	61
W. Europe	1,900	1,900	1,900	586	653	728	1,359	1,516	1,690
Japan-Korea	1,200	1,200	1,200	496	595	713	624	748	897
US	2,100	2,100	2,100	654	736	821	821	900	1,012
Eastern S. America	287	350	426	118	143	175	140	170	207
Western S. America	75	252	252	33	33	33	36	36	43
Central Africa	244	574	574	7	7	7	5	5	5
Other Asia	125	161	196	51	63	76	66	81	98
China	258	291	715	106	106	134	132	149	168
TOTAL	7,933	9,089	10,424	2,784	3,201	3,684	4,107	4,702	5,387

producer of copper concentrates. Zambia also enters in this period, while the Western US shows a modest investment. By 2000, Mexico-Central America and the Western US will realize significant investments in copper mining. It is at this latter period that the exploitation of lower ore grades (mainly Mexico-C. America and Western US) becomes more intense.

Investments on smelters and refineries occur as to match somewhat activity levels for these two processes at each region. However, developing countries tend to export a significant part of their mining output to industrial countries as concentrates. Investments on smelters and refineries situated on developing countries occur mainly to supply nearby semimanufacturing plants. This situation arises to a great extent due to tariff protection on refined copper although transport costs also play a role.

Turning to copper semimanufactures, the following are the main results. As in other scenarios, there are no investments in wire plants at industrial countries. The latter become more dependent on developing countries as their needs for copper wire increase. Thus, Western Europe imports 222,000 tpy of copper wire from Central Africa in 1990 or 10 percent of its requirements, and ends up importing 541,000 tpy from Central Africa and 531,000 tpy from Mexico/C. America or about one-third of its requirements in 2000. Japan-Korea, which are vitually self-sufficient in 1990, import about 15 percent of their consumption by 1995 (91,000 tpy from South Africa and 180,000 tpy from Western South America), and about one-third of their needs in 2000 (30,000 tpy from South Africa, 165,000 tpy from Western South America, and 388,000 tpy from China). USSR-E. Europe, which require to import about 15%

of their needs in 1990 (mostly from the US), import one-third of their market requirements from Mexico-C. America by 2000. Investments in copper wire plants in developing countries depend therefore on two factors: the increasing size of their local markets, as well as on the explosively increasing requirements for imports by industrial countries as their ratio of market requirements to initially installed capacity increases. According to this scenario, by 2000 Mexico-Central America, China, and Central Africa become important producers of copper wire, with productive capacity similar in volume to those of industrial regions.

The case of copper sheets and tubes is very similar to that of previous scenarios. Tariffs protect local markets, labor cost differences are less important, and therefore expansions are almost exclusively in order to supply local markets.

The shadow prices for delivering the last unit of copper to each market may be interpreted as the equilibrium price for copper. The dynamic version is best suited to give an indication of copper prices as it covers three time periods. Table 4.12 shows the shadow prices for selected areas for refined copper and copper semimanfuactures. Differences in prices among the regions are linked to costs of production and transport costs but large variations are mainly related to import tariffs. Analyzing the general trends one would expect copper prices (1981 US dollars) to be about US$1.20/lb in 1990, US$1.40/lb by 1995 and US$1.55/lb in 2000. Thus the increase in copper prices between 1990 and 2000 is 2.6 percent per year in real terms. Since protection is stronger for copper semimanufactures, price differences are in general wider. On the other hand, price/cost escalation for these products

Table 4.12: SHADOW PRICES FOR REFINED COPPER AND SEMIMANUFACTURES

1981 US¢/lb

	1990	1995	2000
Refined Copper			
Mexico/Central America	1.20	1.38	1.55
Western Europe	1.22	1.40	1.57
Japan	1.22	1.38	1.56
United States	1.22	1.39	1.56
Eastern South America	1.35	1.52	1.69
Copper Wire			
Mexico/Central America	1.92	2.10	2.28
Western Europe	2.05	2.23	2.41
Japan	2.14	2.33	2.50
United States	1.94	2.11	2.29
Eastern South America	2.05	2.25	2.41
Copper Sheet and Plate			
Mexico/Central America	1.96	2.08	2.23
Western Europe	2.02	2.16	2.28
Japan	2.10	2.22	2.36
United States	2.02	2.15	2.27
Eastern South America	2.09	2.19	2.35
Copper Tubes and Rods			
Mexico/Central America	1.60	1.74	1.86
Western Europe	1.71	1.83	1.95
Japan	1.68	1.82	1.94
United States	1.64	1.82	1.94
Eastern South America	1.71	1.84	1.96

is not so pronounced as the share of mining costs in these is lower than for refined copper. Thus the price of copper wire escalates at 1.7 percent per year from 1990 to 2000, that of copper sheet and plate at 1.2 percent per year, and copper tubes and rods experience an increase of 1.7 percent per year in the same period.

5. Conclusions

5.1 General Comments

This study presented a framework for analyzing future investment patterns in the copper mining, processing and semifabricating industries in a worldwide context. The mixed integer programming models employed permitted the carrying out of systematic investment analysis taking explicitly into account interdependencies among processing stages, determining the most efficient investment, production and shipment patterns in order to satisfy market requirements.

It could be claimed, that for the level of aggregation presented (a more realistic model would require a higher level of disaggregation), the model handles in an appropriate way the impact of different factors. For example, variation in ore grades has a decisive influence in copper mining, much more important than shipment costs. At the other end of the process, labor costs in semimanufactures seem to determine the location of these activities, although the availability of copper scrap allocates a significant proportion of new investments to industrial countries.

The treatment of the supply side in this model, although lacking some realism due to the almost impossible task of modelling investment behavior, is technically sound as it deals with estimates based on available resources and engineering data. Undoubtedly, this approach represents an advantage versus that of projecting past trends into the future. In order to allow for perception of risk on investments in certain areas (or alternatively cost underestimates in the base data), alternative scenarios considered investment patterns corresponding to higher costs in developing countries.

5.2 Main Trends

The results of this study show that due to the concentration of low cost abundant reserves in a few traditional copper producing areas, regional patterns of production would be mainly unchanged although the relative importance among areas will vary. Thus by 2000, Chile could easily surpass the U.S. and the Soviet Bloc. The fourth producing area in importance could be Mexico and Central America. Zaire, Zambia and Canada would maintain their importance. Together, these seven regions/countries would account for about 85% of world production.

Investments in copper smelters seem to be determined by the result of the trade off between lower investment costs at industrial countries versus lower labor costs at developing countries. In some cases shipment costs weigh the balance of this trade off to one side or the other.

Capacity expansions of copper refineries tend to be located near semimanufacturing plants due to a great extent on tariff protection. If tariffs were to be lifted, as is the case of one scenario, new refinery locations would on average be more distant to semimanufacturing plants although the pattern of the latter would also change.

Wire plants in contrast to sheet/plate and tube plants use mainly primary copper, and besides are the most labor intensive of semimanufacturing plants. Thus, investments in new wire semimanufacturing plants occur in those less developed countries which have easy access to copper supplies and to markets. New sheet/plates and tube semimanufacturing plants occur in both industrial and developing countries. In the former countries scrap availability is an important factor in determining capacity expansion but the trade-off between capital costs and wages is also favorable in some cases to locations in industrial countries. Tariffs on imports of semimanufactures are very important in protecting new investments in Japan-Korea and USSR-E. Europe for sheet/plates and tube plants but they are not sufficient for wire plants.

The dynamic version of the model shows at the mining stage that the traditional copper producers will increase their output levels during the early stages. Only by the end of the period considered, does a relatively new producing area-Mexico-Central America-become important. Investments on smelters and refineries occur as to complement each other. In developing countries these investments are mainly to supply nearby semimanufacturing plants. With respect to semimanufacturing plants, industrial countries increase their dependency on wire imports due to cost differences, but this does not happen on sheet and tubes as tariffs protect local markets, labor cost differences are less important, and copper scrap is abundant in those areas. Finally, the dynamic version offers the time path of marginal costs of production. For refined copper these go from about US$1.20/lb in 1990, to US$1.40/lb by 1995 and US$1.55/lb in 2000 (in 1981 US dollars). Long term price trends for copper should follow this pattern, although productivity improvements could reduce these cost trends.

Notes

1/ See Kendrick and Stoutjesdijk (1975) for a more detailed account of the development of planning models with economies of scale.

2 The description of the processes is based on Bennett et al (1973) and supplemented by other sources as shown.

3/ For a discussion about the origin of such deposits, see Cox et al., (1973).

4/ From Kenji Takeuchi, "Copper Handbook," World Bank, Washington, D.C., 1981.

5/ Prain, Ronald, "Copper - The Anatomy of an Industry," Mining Journal Books Ltd., London, 1975.

6/ Brook Hunt & Associates, Quarterly Service, Issue XII, Copper, London 1982.

Bibliography

Adams, F. Gerard (1973) "The Impact of Copper Production from the Ocean Floor: Application of an Econometric Model," Economics Research Unit, University of Pennsylvania, Philadelphia, Pa.

American Bureau of Metal Statistics (1974). Yearbook of the American Bureau of Metal Statistics.

American Metal Market (1974). Metal Statistics, New York.

Banks, Ferdinand E. (1974). The World Copper Market: An Economic Analysis, Ballinger, Cambridge, Mass.

Baranson, Jack. The Further Processing of Commodities and Raw Materials in Developing Countries for World Markets. Mimeo. World Bank, Washington, D.C.

Barsotti, Aldo F. and Rodney D. Rosenkranz (1983). Natural Resources Forum, United Nations, New York.

Bennett, Harold, Lyman Moore, Lawrence Welborn and Joseph Toland (1973). An Economic Appraisal of the Supply of Copper from Primary Domestic Sources. Bureau of Mines Information Circular, U.S. Department of the Interior, Washington, D.C.

Brook Hunt & Associates, Ltd. (December 1982). Quarterly Service, Issue XII, Copper, London.

Centromin Peru (1975). Esto Es Centromin Peru, Department de Relaciones Publicas de Centromin Peru, Lima, Peru.

CEPAL (Comision Economica Para America Latina) (1969). Los Fletes Maritimos en el Comercio Exterior de America Latina, New York.

Chenery, Hollis B. (January 1952). Overcapacity and the Acceleration Principle. Econometrica, Vol. 20, pp. 1-28.

------, Hollis B. (1959). The Interdependence of Investment Decisions. In the Allocation of Economic Resources edited by M. Abramowitz, Stanford, California.

Chilton, Cecil H. (1960). "Six Tenths Factor Applied to Complete Plant Costs" in C. Chilton (ed.), Cost Engineering in the Process Industries, McGraw-Hill, New York.

CIPEC (1972). Report of the Executive Director to the Governing Board, 1971, Paris, France.

Control Data Corporation (1976). Apex III Reference Manual. Minneapolis, Minnesota.

Cox, Dennis P. et al. (1973). Copper. In: United States Mineral Resources edited by Brobst, Donald A. and Walden P. Pratt. U.S. Department of the Interior, Washington, D.C.

Dammert, Alfredo J. (1977). A World Copper Model for Project Design, Ph.D. dissertation, University of Texas at Austin.

------ (1980). Economia Minera. Universidad del Pacifico, Lima, Peru.

Duloy, John H. and Roger D. Norton (1974). Prices and Incomes in Linear Programmng Models, Development Research Center, I.B.R.D., Washington, D.C.

Economic Report of the President (1976). United States Government Printing Office, Washington, D.C.

Engineering and Mining Journal (January, 1983). E/MJ Survey of Mine and Plant Expansion. McGraw-Hill, New York.

Ffrench Davis, Ricardo and Ernesto Tironi (1974). El Cobre en el Desanollo Nacional. Ediciones Neuva Universidad, Santiago-Chile.

Fisher, Franklin, Paul Cootner and Martin Bailey (Autumn, 1972). "An Econometric Model of the World Copper Industry," Bell Journal of Economics and Management Science, New York.

International Labor Organization (1974). Labor Statistics. Geneva, Switzerland.

International Monetary Fund (monthly). International Financial Statistics, Washington, D.C.

Jalee, Pierre (1968). The Pillage of the Third World. Monthly Review Press, New York.

Kendrick, David (1967). Planning Investments in the Process Industries. M.I.T. Press, Cambridge, Massachusetts.

------, and Ardy Stoutjesdikj (1975). "The Planning of Industrial Investment Programs: A Methodology," Development Research Center, I.B.R.D., Washington, D.C.

Manne, Alan S. (editor, 1967). Investments for Capacity Expansion: Size, Location and Time-Phasing. George Alland Unwin Ltd., London.

Manners, Gerald (1971). The Changing World Markets for Iron Ore 1950-1980. John Hopkins, Baltimore, Md.

Markowitz, H.M. and A.S. Manne (January 1957). On the Solution of Discrete Programming Problems. Econometrica, Vol. 25, No. 1.

McNicol, D.L. (1975). "The Two-Price System in the Copper Industry," The Bell Journal of Economics, Vol. 6, No. 1, New York.

Meeraus, A., A. Stoutjesdijk and D. Weigel (1975). "An Investment Planning Model for the World Fertilizer Industry," I.B.R.D./IFC, Washington, D.C. Metal Bulletin Limited (1974). Copper 1974, London.

Metallgesellschaft Aktiengesellschaft (annual). Metal Statistics, Frankfurt am Main: Metallgesellschaft Aktiengesellschaft, Germany.

Mingst, Karen (Spring 1976). Cooperation or Illusion. International Organization, Vol. 30, No. 2. Stanford University, California.

Moran, T.H. and D.H. Maddox (1980). Structure and Strategy in the International Copper Industry, (draft). United Nations Centre for Transnational Corporations, New York.

Morrison, Thomas K. (1976). Manufactured Exports from Developing Countries. Praeger, New York.

OECD (1973, 1974). The Non Ferrous Metal Industry. Organization for Economic Cooperation and Development, Paris, France.

Oxford Economic Atlas of the World (1972). Oxford University Press, London.

Prain, Ronald (1975). Copper - The Anatomy of an Industry. Mining Journal Books, Ltd., London.

Radetzki, Marian (February-March, 1975). Metal Mineral Resource Exhaustion and the Threat to Material Progress - The Case of Copper. World Development, Vol. 3, Nos. 2 and 3, Cambridge, England.

Rodriguez, Hoyle, ed. (1974). Peru Minero 1974. Sociedad Nacional de Mineria y Petroleo, Lima, Peru.

Sutulov, Alexander (ed.) (1975). El Cobre Chileno. Corporacion del Cobre, Santiago, Chile.

Takayama, T. and J.G. Judge (1971). Spatial and Temporal Price and Allocation Models, North Holland Publishing Co., Amsterdam.

Takeuchi, Kenji (1975). Copper: Markets Prospects for 1980 and 1985. In Natural Resources and National Welfare - The Case of Copper, edited by Ann Seidman. Praeger, New York.

------, Kenji (1981). Copper Handbook, World Bank, Washington, D.C.

Underwood, John (1976). Optimizing Rules for Producer Groups in a Stochastic Market Setting with Applications to the Copper and Tea Markets. Ph.D. dissertation, University of Minnesota.

UNCTAD (1982). Processing and Marketing of Copper: Areas for International Cooperation, Geneva.

United Nations (annual). Statistical Yearbook, New York.

-------, (annual). Yearbook of National Account Statistics, New York.

United Nations Industrial Development Organization (1972). Copper Production in Developing Countries, New York.

------, (1972). Non Ferrous Metals - A Survey of Their Production and Potential in the Developing Countries, New York.

United Nations Statistical Office (196401972). World Trade Annual and Supplements, Walker and Company, New York.

Vietorisz, T., and A.S. Manne (1963). Chemical Processes, Plant Location, and Economies of Scale. In studies in Process Analysis edited by A.S. Manne and H.M. Markowitz. John Wiley and Sons, New York.

Westphal, Larry (1971). Planning Investments With Economies of Scale. North-Holland, Amsterdam.

Wimpfen, Sheldon P. and Harold J. Bennett (1975). "Copper Resource Appraisal," paper presented at 1973 National Research Council Committee on Mineral Resources and the Environment (COMRATE). Panel III meeting on Mineral Resource Appraisals and Copper Resources, Estes Park, Colorado and in Resources Policy, March, 1975.

Woods, Donald R. (1975). Financial Decision Making in the Process Industry. Prentice Hall, New Jersey.

APPENDIX
Computer-Readable Representation of the Model

```
GAMS 1.208  MODELING INVESTMENTS IN THE WORLD COPPER SECTOR                       02/16/84    14.28.26.   PAGE    1
            PROBLEMS AND NOTES

                                                                                              NEW MARGIN = 002-120
   4   * SCENARIO DEFINITIONS:
   5   *
   6   *   BASE CASE:  WITH TARIFFS, MINIMUM WIRE PRODUCTION LEVEL (AT INSTALLED CAPACITY) FOR YEAR 2000
   7   *
   8   *   EXPT. 1   :  NO TARIFFS
   9   *
  10   *   EXPT. 2   :  REDUCE OPERATING COST OF MEDIUM LEVEL ORE PROCESSING IN CHILE, PERU, AND USSR BY 25%
  11   *
  12   *   EXPT. 3   :  NO MINIMUM WIRE PRODUCTION
  13   *
  14   *   EXPT. 4   :  INCREASE OPERATING COSTS IN LDCS BY 25%
  15   *
  16   *   EXPT. 5   :  INCREASE INVESTMENT COSTS IN LDCS BY 25%
  17   *
  18   *   EXPT. 6   :  DYNAMIC VERSION OF BASE CASE WITH 5 YEAR PERIODS - 1985, 1990, 1995, 2000.


GAMS 1.208  MODELING INVESTMENTS IN THE WORLD COPPER SECTOR                       02/16/84    14.28.26.   PAGE    2
            SET DEFINITIONS

  20    SET IRUN     MINE SMELTER AND REFINERY LOCATIONS /
  21
  22                 PERU       , CHILE      , ZAMBIA , ZAIRE     , MEX+CAM    , S-AFRICA  , PHILIPINES , PAPUA-NG
  23                 WESTERN-US , EASTERN-US, CANADA , EE+USSR   , AUSTRALIA , W-EUROPE  , JAPAN+KOR  /
  24
  25        I(IRUN)  MINE SMELTER AND REFINERY LOCATION /
  26
  27                 PERU       , CHILE      , ZAMBIA , ZAIRE     , MEX+CAM    , S-AFRICA  , PHILIPINES , PAPUA-NG
  28                 WESTERN-US , EASTERN-US, CANADA , EE+USSR   , AUSTRALIA , W-EUROPE  , JAPAN+KOR  /
  29
  30        JRUN     WIRE TUBE AND SHEET PLANT AND MARKET LOCATIONS /
  31                 USA       , MEX+CAM    , ES-AMERICA , WS-AMERICA  , W-EUROPE   , EE+USSR
  32                 C-AFRICA , S-AFRICA   , O-ASIA      , JAPAN+KOR   , CHINA      , AUSTRALIA , CANADA    /
  33
  34        J(JRUN)  WIRE TUBE AND SHEET PLANT AND MARKET LOCATIONS /
  35                 USA       , MEX+CAM    , ES-AMERICA , WS-AMERICA  , W-EUROPE   , EE+USSR
  36                 C-AFRICA , S-AFRICA   , O-ASIA      , JAPAN+KOR   , CHINA      , AUSTRALIA , CANADA    /
  37
  38        C        COMMODITIES /
  39
  40                 ORE
  41                 SCRAP-S          SCRAP COPPER FOR SMELTING
  42                 BLISTER
  43                 SCRAP-R          SCRAP COPPER FOR REFINING
  44                 REFINED-CU       REFINED COPPER
  45                 SCRAP-SPS        SCRAP COPPER FOR SHEETS PLATES AND STRIP FABRICATION
  46                 SHEETS+P+S       SHEETS PLATES AND STRIP
  47                 WIRE
  48                 SCRAP-T          SCRAP COPPER FOR TUBES AND RODS FABRICATION
  49                 TUBES+RODS       TUBES AND RODS                                /
  50
  51        CM(C)    COMMODITIES IN MINING AND PROCESSING     / ORE, SCRAP-S, BLISTER, SCRAP-R, REFINED-CU /
  52
  53        CS(C)    COMMODITIES AT WIRE TUBE & SHEET PLANTS / REFINED-CU, WIRE, SCRAP-T, TUBES+RODS, SCRAP-SPS, SHEETS+P+S /
  54
  55        CF(C)    FINAL PRODUCTS                           / REFINED-CU, WIRE, TUBES+RODS, SHEETS+P+S /
  56
  57        CFR(C)   FINAL PRODUCTS FROM PROCESSING           / REFINED-CU /
  58
  59        CFS(C)   FINAL PRODUCTS FROM SEMI-MANUFACTURE     / WIRE, TUBES+RODS, SHEETS+P+S /
  60
  61        CIM(C)   FINAL PRODUCTS AT MINES AND SMELTERS     / ORE, BLISTER /
  62
  63        CIL(C)   SCRAP TYPES                              / SCRAP-S, SCRAP-R, SCRAP-T, SCRAP-SPS /
  64
  65        P        PROCESSES /
  66
  67                 HIGH-GRADE     HIGH GRADE ORE MINING BY OPEN PIT AND CONCENTRATOR
  68                 MED-GRADE      MEDIUM GRADE ORE MINING BY OPEN PIT AND CONCENTRATOR
  69                 SMELTING-O     SMELTING USING PRIMARY COPPER ORE
  70                 SMELTING-S     SMELTING USING SCRAP COPPER
  71                 REFINING-B     REFINING USING BLISTER
  72                 REFINING-S     REFINING USING SCRAP COPPER
  73                 WIRE-REF-C     WIRE FABRICATION FROM REFINED COPPER
  74                 TUBE-REF-C     TUBE AND ROD FABRICATION FROM REFINED COPPER
  75                 TUBE-SCRAP     TUBE AND ROD FABRICATION FROM SCRAP COPPER
```

```
GAMS 1.208  MODELING INVESTMENTS IN THE WORLD COPPER SECTOR                                  02/16/84     14.28.26.    PAGE     3
            SET DEFINITIONS

  76                  S-REF-C       SHEET PLATE AND STRIP FABRICATION FROM REFINED COPPER
  77                  S-SCRAP       SHEET PLATE AND STRIP FABRICATION FROM SCRAP COPPER  /
  78
  79         PM(P)    PROCESS AT MINES SMELTERS AND REFINERIES   / HIGH-GRADE, MED-GRADE  , SMELTING-O
  80                                                               SMELTING-S, REFINING-B , REFINING-S   /
  81
  82         PMH(P)   HIGH GRADE ORE MINING PROCESSES            / HIGH-GRADE , MED-GRADE /
  83
  84         PMM(P)   MINING PROCESSES                           / HIGH-GRADE, MED-GRADE /
  85
  86         PS(P)    SMELTING PROCESSES                         / SMELTING-O  , SMELTING-S /
  87
  88         PR(P)    REFINING PROCESSES                         / REFINING-B  , REFINING-S /
  89
  90         PSM(P)   SEMI-MANUFACTURING PROCESS
  91
  92         PW(P)    WIRE PLANT PROCESSES                       / WIRE-REF-C /
  93
  94         PTR(P)   TUBE AND ROD PLANT PROCESSES               / TUBE-REF-C, TUBE-SCRAP /
  95
  96         PSH(P)   SHEET PLANT PROCESSES                      / S-REF-C, S-SCRAP /
  97
  98         M        PRODUCTIVE UNITS /
  99
 100                  OPEN-PIT      OPEN PIT MINES AND CONCENTRATORS
 101                  SMELTER
 102                  REFINERY
 103                  WIRE          WIRE FABRICATION PLANT
 104                  TUBES+RODS    TUBE AND ROD FABRICATION PLANT
 105                  SHEETS+P+S    SHEET PLATE AND STRIP FABRICATION PLANT /
 106
 107         MM(M)    PRODUCTIVE UNITS AT MINING AND PROCESSING PLANTS / OPEN-PIT, SMELTER, REFINERY/
 108
 109         MS(M)    PRODUCTIVE UNITS AT SEMI-MANUFACTURE PLANTS
 110
 111         TG       TIME PERIODS          /  1980*2000  /
 112
 113         T(TG)    SOLUTION TIME PERIOD / 2000 /
 114
 115         TBASE(TG)                      / 1980 /
 116
 117
 118     ALIAS (TG,TGP),(I,IP),(J,JP),(T,TP),(IRUN,IRUNP),(JRUN,JRUNP);
 119     PSM(P) = NOT PM(P); MS(M) = NOT MM(M);
 120
 121     DISPLAY PSM,PM;
```

```
GAMS 1.208  MODELING INVESTMENTS IN THE WORLD COPPER SECTOR                                  02/16/84     14.28.26.    PAGE     4
            SET DEFINITIONS

 123   * TRANSFORM THE SET OF YEAR LABELS TO NUMERICAL VALUES AND COMPUTE LENGTH OF TIME PERIODS USED IN MODEL:
 124
 125     PARAMETERS YBASE(TG)  DISTANCE OF YEAR FROM BASE YEAR
 126                PERIOD(TG) INTERVAL LENGTH (USED IN OBJECTIVE FUNCTION)
 127                RPH(TG)    LENGTH OF PERIODS FOR RESERVES;
 128
 129     YBASE(TG) = ORD(TG) - SMIN(TGP$TBASE(TGP), ORD(TGP));
 130     PERIOD(T) = YBASE(T) - YBASE(T-1);
 131     RPH(T)    = PERIOD(T);
 132
 133     DISPLAY YBASE,PERIOD,RPH;
 134
 135   * FOR THE STATIC MODEL WE DON'T NEED TO HAVE THE PERIOD LENGTH INFORMATION.  THEREFORE RESET PERIOD TO 1.
 136
 137     PERIOD(T) = 1;
 138
 139     SCALAR    LBPERTON        CONVERSION FROM LBS TO TONS   / 2240 /;
```

```
GAMS 1.208  MODELING INVESTMENTS IN THE WORLD COPPER SECTOR                                02/16/84    14.28.26.   PAGE    5
            DEMAND DATA

   141     SET CGR          CONSUMPTION GROWTH RATE INTERVALS  / G1, G2 /
   142         TG1(TG)      INTERVAL 1 DEFINITION              / 1980*1985 /
   143         TG2(TG)      INTERVAL 2 DEFINITION              / 1986*2000 /
   144         MGR(CGR,TG)  MAP OF INTERVAL TO YEARS           / G1.(1980*1985) , G2.(1986*2000) /
   145
   146     PARAMETER DEMAND(JRUN,CF,TG)  DEMAND FOR REFINED COPPER AND SEMI-MANUFACTURES (1000 TONS)
   147
   148
   149     TABLE DEM1980(JRUN,*)  CONSUMPTION IN 1980 AND GROWTH RATES (1000 TONS & %): TABLE
   150
   151                                 WIRE     TUBES+RODS    SHEETS+P+S    REFINED-CU      G1        G2
   152    *                                                                             (1980-85) (1986-2000)
   153
   154     USA                         1385         690          555           40          1.3       2.2
   155     MEX+CAM                       80          40           30            4          3.9       4
   156     ES-AMERICA                   195          95           80           16          3.9       4
   157     WS-AMERICA                    40          20           15            3          3.9       4
   158     W-EUROPE                    2075        1160          500          205          1         2.2
   159     EE+USSR                     1050         525          420           55          2.9       3
   160     C-AFRICA                      15           /            5            1          3.9       4
   161     S-AFRICA                      55          30           20            3          3.9       4
   162     O-ASIA                        90          45           35            6          3.9       4
   163     JAPAN+KOR                    875         440          350           45          3.4       3.7
   164     CHINA                        195         100           80           11          3.3       2.4
   165     AUSTRALIA                     85          40           35            6          1.3       2.4
   166     CANADA                       125          60           50            5          1.3       2.4
   167                                                                                                ;
   168
   169
   170     DEMAND(JRUN,CF,TG1)   = DEM1980(JRUN,CF) * (1 + SUM(CGR$MGR(CGR,TG1), DEM1980(JRUN,CGR))/100)**(ORD(TG1) - 1);
   171     DEMAND(JRUN,CF,TG2)   = DEMAND(JRUN,CF,"1985") * (1 + SUM(CGR$MGR(CGR,TG2), DEM1980(JRUN,CGR))/100)**ORD(TG2);
   172
   173     DISPLAY DEMAND;
```

```
GAMS 1.208  MODELING INVESTMENTS IN THE WORLD COPPER SECTOR                                02/16/84    14.28.26.   PAGE    6
            TECHNOLOGY DATA

   175     TABLE A(C,P)  INPUT-OUTPUT COEFFICIENTS
   176
   177                  HIGH-GRADE   MED-GRADE    SMELTING-O   SMELTING-S   REFINING-B   REFINING-S
   178
   179     ORE              1.           1.          -1.03
   180     BLISTER                                    1.          1.          -1.03
   181     REFINED-CU                                                           1.           1.
   182     SCRAP-S                                               -1.03
   183     SCRAP-R                                                                         -1.03
   184
   185        +         WIRE-REF-C   TUBE-REF-C   TUBE-SCRAP   S-REF-C   S-SCRAP
   186
   187     REFINED-CU   -1.007        -.718                     -.772
   188     SCRAP-T                                  -.718
   189     SCRAP-SPS                                                      -.772
   190     WIRE          1.
   191     TUBES+RODS                   1.           1.
   192     SHEETS+P+S                                            1.        1.
   193
   194
   195
   196     TABLE B(M,P)  CAPACITY UTILIZATION MATRIX
   197
   198                  HIGH-GRADE   MED-GRADE   SMELTING-O   SMELTING-S   REFINING-B   REFINING-S
   199
   200     OPEN-PIT         1.          1.6
   201     SMELTER                                  1.          .32
   202     REFINERY                                                          1.           1.
   203
   204        +         WIRE-REF-C   TUBE-REF-C   TUBE-SCRAP   S-REF-C   S-SCRAP
   205
   206     WIRE          1.
   207     TUBES+RODS                   1.           1.
   208     SHEETS+P+S                                            1.        1.
   209
   210
   211    *
   212    * IN THE BASE CASE MINIMUM WIRE PRODUCTION LEVEL IS SET TO THE INSTALLED CAPACITY.
```

```
GAMS 1.208  MODELING INVESTMENTS IN THE WORLD COPPER SECTOR                              02/16/84     14.28.26.    PAGE     7
            SCRAP DATA

   214    * 1. SCRAP COST IS ADDED TO THE "BASE" PROCESS OPERATING COST TO GIVE THE OPERATING COST OF PROCESSING SCRAP
   215    *    (SEE OPERATING COST SECTION).  THE MAJOR COMPONENT OF SCRAP COST IS COLLECTION.
   216    *
   217    * 2. SCRAP QUALITY DETERMINES IF IT WILL BE USED IN SMELTING, REFINING OR SEMI-MANUFACTURING.
   218    *
   219    * 3. 1980 USE OF SCRAP IN REFINERIES & SMELTERS IS KNOWN, AND IS EXPECTED TO GROW AT A CONSTANT RATE.
   220    *
   221    * 4. IN THE MODEL WE CAN RESTRICT THE SCRAP AVAILABILITY FOR SEMI-MANUFACTURING AT EITHER THE AGGREGATE
   222    *    SEMI-MANUFACTURING LEVEL OR AT THE PARTICULAR PROCESS (TUBE AND RODS, AND SHEETS) LEVEL.
   223
   224     PARAMETER SCRAPI(IRUN,TG,C) SCRAP AVAILABILITY AT REFINERIES AND SMELTERS (1000 TONS)
   225               SCRAPJ(JRUN,TG,*) SCRAP AVAILABILITY AT SEMI-MANUFACTURERS      (1000 TONS);
   226
   227
   228     TABLE HDS1(IRUN,*) SCRAP USAGE IN 1980 AT SMELTERS AND REFINERIES (1000 TONS)
   229
   230                    SCRAP-S   SCRAP-R    GROWTH
   231    *                                     (%)
   232
   233     WESTERN-US       48         193.6       4
   234     EASTERN-US       12         246.4       4
   235     W-EUROPE        130         460         1.7
   236     JAPAN+KOR        50         141         4
   237     AUSTRALIA         5          28         1.1
   238     CANADA                       40         4
   239     MEX+CAM                       5         4
   240
   241
   242     TABLE HDS2(JRUN,*)   SCRAP USAGE IN SEMI-MANUFACTURING IN 1980 (1000 TONS)
   243
   244                    SCRAP-T      SCRAP-SPS   OTHER     GROWTH
   245    *                                                    (%)
   246
   247     USA               440           340        100      1
   248     W-EUROPE          490           210        220      3.5
   249     JAPAN+KOR         230           140         30      2.5
   250     CANADA              5             4         10      3.5
   251     AUSTRALIA          21            18          4      1.1
   252                                                              ;
   253
   254     SCRAPI(IRUN,TG,CIL)        = HDS1(IRUN,CIL) * (1 + HDS1(IRUN,"GROWTH")/100)**(ORD(TG)-1);
   255     SCRAPJ(JRUN,TG,CIL)        = HDS2(JRUN,CIL) * (1 + HDS2(JRUN,"GROWTH")/100)**(ORD(TG)-1);
   256     SCRAPJ(JRUN,TG,"SCRAP")    = SUM(CIL, SCRAPJ(JRUN,TG,CIL));
   257
   258     DISPLAY SCRAPI,SCRAPJ;
   259
   260     SET       PSI(P)        REFINERY AND SMELTER PROCESSES USING SCRAP COPPER INPUT
   261               PSJ(P)        WIRE TUBE AND SHEET PROCESSES USING SCRAP COPPER INPUT
   262
   263     PARAMETER MAPIC(IRUN,C) COMMODITIES AT MINES REFINERIES AND SMELTERS
   264               MAPIP(IRUN,P) PROCESSES AT MINES REFINERIES AND SMELTERS
   265               MAPJC(JRUN,C) COMMODITIES AT WIRE TUBE AND SHEET PLANTS
   266               MAPJP(JRUN,P) PROCESSES AR WIRE TUBE AND SHEET PLANTS;
```

```
GAMS 1.208  MODELING INVESTMENTS IN THE WORLD COPPER SECTOR                              02/16/84     14.28.26.    PAGE     8
            SCRAP DATA

   268    * THE SUBSET OF SCRAP USING PROCESSES IS IDENTIFIED AS ONLY THOSE PROCESSES WHICH UTILIZE SCRAP MATERIALS AS INPUTS:
   269
   270     PSI(PM)  = YES$(SUM(CIL, A(CIL,PM)) LT 0);
   271     PSJ(PSM) = YES$(SUM(CIL, A(CIL,PSM)) LT 0);
   272
   273    * FIRST DEFINE ALL COMMODITIES TO BE AVAILABLE AT ALL LOCATIONS.  THEN, IF NO SCRAP IS AVAILABLE AT A LOCATION
   274    * THIS SCRAP TYPE IS THEN DEFINED TO NOT EXIST AT THE GIVEN LOCATION.  SIMILARLY, FIRST ALL PROCESSES ARE DEFINED
   275    * TO EXIST AT ALL LOCATIONS; IF THE SCRAP COMMODITY IS NOT AVAILABLE AT A LOCATION AND IF A PROCESS USES THIS
   276    * SCRAP COMMODITY AS INPUT, THEN THIS PROCESS IS DEFINED TO NOT EXIST AT THE LOCATION.
   277
   278     MAPIC(IRUN,CM)                          = 1;
   279     MAPIC(IRUN,CIL)$(HDS1(IRUN,CIL) EQ 0) = 0;
   280     MAPIP(IRUN,PM)                          = 1;
   281     MAPIP(IRUN,PSI)$(SUM(CIL, A(CIL,PSI)*MAPIC(IRUN,CIL)) EQ 0) = 0;
   282
   283     MAPJC(JRUN,CS)                          = 1;
   284     MAPJC(JRUN,CIL)$(HDS2(JRUN,CIL) EQ 0) = 0;
   285     MAPJP(JRUN,PSM)                         = 1;
   286     MAPJP(JRUN,PSJ)$(SUM(CIL, A(CIL,PSJ)*MAPJC(JRUN,CIL)) EQ 0) = 0;
   287
   288     DISPLAY MAPIC,MAPIP,MAPJC,MAPJP;
```

```
 290      PARAMETER RESC(IRUN,P) ANNUAL EXTRACTION LIMITS ON MEDIUM AND HIGH GRADE ORES (THOUSAND TONS)
 291
 292      TABLE RESERVES(IRUN,P)  ORE RESERVES ESITMATES FOR NEXT 20 YEARS (1980-81 - MILLION TONS)
 293
 294                                        HIGH-GRADE     MED-GRADE
 295
 296      PERU                                    1.67        18.8
 297      CHILE                                  55.77        17.29
 298      ZAMBIA                                 18.01         1.21
 299      ZAIRE                                  16.97         1.48
 300      MEX+CAM                                 3.61        20.14
 301      S-AFRICA                                2.42          .76
 302      PAPUA-NG                                             9.5
 303      PHILIPINES                              4.93         3.97
 304      WESTERN-US                             19.38        25.87
 305      EASTERN-US                                           5.35
 306      CANADA                                  8.33         9.6
 307      EE+USSR                                38           17
 308      AUSTRALIA                               6              .8
 309      W-EUROPE                                             7.87
 310      JAPAN+KOR                                            1.4
 311                                                                                ;
 312
 313      RESC(IRUN,PMH) = 1000*RESERVES(IRUN,PMH)/20;
 314      DISPLAY RESC;
 315
 316
 317      TABLE CAPS(JRUN,M)  EXISTING CAPACITY IN SEMI-MANUFACTURING IN 1980 (1000 TONS)
 318
 319                          WIRE         TUBES+RODS         SHEETS+P+S
 320
 321      USA                2100             900                1000
 322      MEX+CAM             105              40                  15
 323      ES-AMERICA          270              65                  30
 324      WS-AMERICA           55              15                   9
 325      W-EUROPE           1900            1200                 900
 326      EE+USSR            1200             600                 350
 327      C-AFRICA             15               5                   7
 328      S-AFRICA             55              40                  25
 329      O-ASIA              125              30                   4
 330      JAPAN+KOR          1200             600                 410
 331      CHINA               250              85                  60
 332      AUSTRALIA           130              70                  70
 333      CANADA              200              50                  60
```



```
 335      TABLE CAPM(IRUN,M)  ESTIMATES OF MINE SMELTER AND REFINERY CAPACITIES IN 1980 (1000 TONS)
 336
 337                            OPEN-PIT        SMELTER       REFINERY
 338
 339      PERU                     590             350           230
 340      CHILE                   1330             950           810
 341      ZAMBIA                   550             601           610
 342      ZAIRE                    630             430           140
 343      MEX+CAM                  324              86           100
 344      S-AFRICA                 490             280           150
 345      PHILIPINES               492
 346      PAPUA-NG                 256
 347      WESTERN-US              1675            1510           480
 348      EASTERN-US               576             540          1670
 349      CANADA                  1254             610           610
 350      EE+USSR                 1720            1720          1600
 351      AUSTRALIA                260             220           210
 352      W-EUROPE                 608             820          1440
 353      JAPAN+KOR                112            1180          1220
```

```
GAMS 1.208  MODELING INVESTMENTS IN THE WORLD COPPER SECTOR                                02/16/84    14.28.26.   PAGE   11
            INVESTMENT DATA

  355    SET IH(IRUN)      REFINERIES & SMELTERS WITH HIGH COST         / PERU     , CHILE     , MEX+CAM   , ZAMBIA
  356                                                                    ZAIRE    , S-AFRICA , PAPUA-NG , PHILIPINES /
  357        IL(IRUN)      REFINERIES & SMELTERS WITH LOW COST
  358        JH(JRUN)      SEMI-MANUFACTURING LOCATIONS WITH HIGH COST / MEX+CAM   , ES-AMERICA, WS-AMERICA
  359                                                                    C-AFRICA , S-AFRICA  , O-ASIA      , CHINA     /
  360        JL(JRUN)      SEMI-MANUFACTURING LOCATIONS WITH LOW COST
  361
  362        IT            COPPER PROCESSING CLASSIFICTION              / IIH   REQUIRING HIGH COST, IIL   REQUIRING LOW COST /
  363        IIT(IT,IRUN)  MAP OF LOCATIONS TO CLASSIFICATION
  364        JT            SEMI-MANUFACTURING CLASSIFICATION            / INDUSTRIAL, NON-IND /
  365        JTJ(JT,JRUN)  MAP OF LOCATIONS TO CLASSIFICATION;
  366
  367    IL(IRUN) = NOT IH(IRUN);
  368    JL(JRUN) = NOT JH(JRUN);
  369
  370    IIT("IIH",IH)          = YES;  IIT("IIL",IL)      = YES;
  371    JTJ("INDUSTRIAL",JL) = YES;  JTJ("NON-IND",JH) = YES;
  372
  373    DISPLAY IL,JL,IIT,JTJ;
  374
  375
  376    SCALAR     RHO      DISCOUNT RATE           / .1 /
  377               LIFE     LIFE OF UNITS    (YRS)    / 20 /
  378
  379    PARAMETER DIS       DISCOUNT FACTOR
  380              SIGMA     CAPITAL RECOVERY FACTOR
  381
  382              OMEGAM    SCALE COST OF MINING AND PROCESSING PLANTS                   (MILLION US$ PER 1000 TPY)
  383              OMEGAS    SCALE COST OF OF SEMI-MANUACTURING PLANTS                    (MILLION US$ PER 1000 TPY)
  384              NUM       PROPORTIONAL CAPITAL COST OF MINING AND PROCESSING PLANTS (MILLION US$ PER 1000 TPY)
  385              NUS       PROPORTIONAL CAPITAL COST OF SEMI-MANUF PLANTS               (MILLION US$ PER 1000 TPY)
  386
  387              HBARM     ECONOMIES OF SCALE SIZE FOR PROCESSING PLANTS (1000 TONS)
  388              HBARS     ECONOMIES OF SCALE SIZE FOR SEMI-MANUF PLANTS (1000 TONS)
  389              HHATM     MAXIMUM SIZES FOR PROCESSING PLANTS            (1000 TONS)
  390              HHATS     MAXIMUM SIZES FOR SEMI-MANUFACTURING PLANTS   (1000 TONS)
  391
  392              TS(T,TP)  TIME SUMMATION MATRIX
  393              SFI(IRUN) SITE FACTOR FOR MINING AND PROCESSING PLANTS
  394              SFJ(JRUN) SITE FACTOR FOR SEMI-MANUFACTURING PLANTS;
  395
  396
  397    TABLE INVM(IRUN,*)  INVESTMENT DATA FOR MINES - 1980(1)
  398
  399                                              FIXED           VAR          EC-SCALE
  400   *                                       (MILL US$)   (MILL US$/1000 TON)(1000 TONS)
  401
  402    (PERU,CHILE,MEX+CAM,S-AFRICA
  403     PHILIPINES,PAPUA-NG)                      40             5.6             100
  404    (ZAMBIA,ZAIRE)                           120             5.8             380
  405    (WESTERN-US,CANADA,EE+USSR,AUSTRALIA
  406     EASTERN-US,W-EUROPE,JAPAN+KOR)            40             5.06             75


GAMS 1.208  MODELING INVESTMENTS IN THE WORLD COPPER SECTOR                                02/16/84    14.28.26.   PAGE   12
            INVESTMENT DATA

  408    TABLE INVS(JT,M,*)  INVESTMENT COSTS FOR SEMI-MANUFACTURING - 1980
  409
  410                                          FIXED           VAR              EC-SCALE
  411   *                                    (MILL US$)  (MILL US$/1000 TONS)    (1000 TONS)
  412
  413    (INDUSTRIAL).WIRE                     3.38          1.72                20
  414    (NON-IND    ).WIRE                    3.6           1.84                20
  415    (INDUSTRIAL).TUBES+RODS               5.64           .94                30
  416    (NON-IND    ).TUBES+RODS              6             1                   30
  417    (INDUSTRIAL).SHEETS+P+S               8.64          1.34                30
  418    (NON-IND    ).SHEETS+P+S              9.2           1.44                30
  419
  420
  421    TABLE INVRS(IT,M,*)  INVESTMENT COSTS FOR REFINERY AND SMELTER - 1980(1)
  422
  423                                            FIXED            VAR               EC-SCALE
  424   *                                      (MILL US$) (MILL US$/1000 TONS)   (1000 TONS)
  425
  426    IIH.SMELTER                               50             2.166              150
  427    IIL.SMELTER                               40             1.832              150
  428    IIH.REFINERY                               5.02           .718              150
  429    IIL.REFINERY                               4.02           .574              150
  430                                                                                    ;
  431
  432
  433    HBARM(IRUN,M)              = SUM(IT$IIT(IT,IRUN), INVRS(IT,M,"EC-SCALE"));
  434    HBARM(IRUN,"OPEN-PIT")     = INVM(IRUN,"EC-SCALE");
  435    HBARS(JRUN,M)              = SUM(JT$JTJ(JT,JRUN), INVS(JT,M,"EC-SCALE"));
  436
```

```
437     HHATM(IRUN,M)              = 40*HBARM(IRUN,M);
438     HHATS(JRUN,M)              = 40*HBARS(JRUN,M);
439
440     OMEGAM(IRUN,M)$HBARM(IRUN,M)                  = SUM(IT$IIT(IT,IRUN), INVRS(IT,M,"FIXED"))/HBARM(IRUN,M);
441     OMEGAM(IRUN,"OPEN-PIT")$HBARM(IRUN,"OPEN-PIT") = INVM(IRUN,"FIXED")/HBARM(IRUN,"OPEN-PIT");
442     OMEGAS(JRUN,M)$HBARS(JRUN,M)                  = SUM(JT$JTJ(JT,JRUN), INVS(JT,M,"FIXED"))/HBARS(JRUN,M);
443
444     NUM(IRUN,M)                 = OMEGAM(IRUN,M) + SUM(IT$IIT(IT,IRUN), INVRS(IT,M,"VAR"));
445     NUM(IRUN,"OPEN-PIT")        = OMEGAM(IRUN,"OPEN-PIT") + INVM(IRUN,"VAR");
446     NUS(JRUN,M)                 = OMEGAS(JRUN,M) + SUM(JT$JTJ(JT,JRUN), INVS(JT,M,"VAR"));
447
448     SFI(IRUN)  = 1;
449     SFJ(JRUN)  = 1;
450
451     DIS(T) = (1 + RHO)**(-YBASE(T));
452     DIS(T) = 1;
453     SIGMA = RHO*(1 + RHO)**LIFE / ((1 + RHO)**LIFE - 1);
454
455     TS(T,TP) = 1$(ORD(T) GE ORD(TP));
456     DISPLAY HBARM,HBARS,OMEGAM,OMEGAS,NUM,NUS,TS,DIS,SIGMA;
```



```
458     PARAMETER OPM(IRUN,P)  MINING AND PROCESSING COSTS IN 1980(1) (US$ PER TON OF COPPER CONTENT)
459
460             SPR          SCAP PRICES (AT COPPER PRICE OF US$2000 A TON) / SEMI        1800   , SMELTING-S  1200
461                                                                          REFINING-S  1700                   /
462
463
464             OM(IRUN)     MINE OPERATING COST IN 1980(1)  (US$ PER TON OF COPPER CONTENT)   /
465
466                          (PERU,MEX+CAM,S-AFRICA)                                              840
467                          CHILE                                                               1028
468                          ZAMBIA                                                              1044
469                          ZAIRE                                                                681
470                          PHILIPINES                                                          1054
471                          PAPUA-NG                                                             295
472                          (WESTERN-US,EASTERN-US)                                              771
473                          (CANADA,EE+USSR,AUSTRALIA,W-EUROPE,JAPAN+KOR)                         914      /
474
475             OS(IRUN)     SMELTER OPERATING COST IN 1980(1) (US$ PER TON OF COPPER CONTENT) /
476
477                          (PERU,MEX+CAM,S-AFRICA,PHILIPINES,PAPUA-NG,CHILE,ZAMBIA,ZAIRE)       250
478                          (WESTERN-US,EASTERN-US,CANADA,JAPAN+KOR,AUSTRALIA,W-EUROPE,EE+USSR)   300      /
479
480             OPR(IRUN)    REFINERY OPERATING COST IN 1980(1) (US$ PER TON OF COPPER CONTENT)/
481
482                          (PERU,MEX+CAM,S-AFRICA,PHILIPINES,PAPUA-NG,CHILE,ZAMBIA,ZAIRE)       150
483                          (WESTERN-US,EASTERN-US,CANADA,JAPAN+KOR,AUSTRALIA,W-EUROPE,EE+USSR)   170      /
484
485              OPI(IRUN)   OPERATING COST ESCALATOR FOR MINING AND PROCESSING LOCATIONS
486
487              OPJ(JRUN)   OPERATING COST ESCALATOR FOR SEMI-MANUFACTURING LOCATIONS
488
489
490     TABLE OPS(JRUN,P)  OPERATING COSTS FOR SEMI-MANUFACTURING IN 1980(1) (1000 US$ PER TON)
491
492                                            WIRE-REF-C  TUBE-REF-C     S-REF-C
493     (USA,W-EUROPE,EE+USSR,
494      JAPAN+KOR,AUSTRALIA,CANADA)              1.87        1.687         2.36
495     (MEX+CAM,ES-AMERICA,
496      WS-AMERICA,C-AFRICA,S-AFRICA)            1.33        1.507         2.09
497     (CHINA,O-ASIA)                            1.15        1.447         2
498                                               ;
499
500
501     OPM(IRUN,PMM)$MAPIP(IRUN,PMM) = OM(IRUN);
502     OPM(IRUN,PS)$MAPIP(IRUN,PS)   = OS(IRUN);
503     OPM(IRUN,PR)$MAPIP(IRUN,PR)   = OPR(IRUN);
504
505    * THE OPERATING COST FOR LOCATIONS WITH SCRAP PROCESS ARE ADJUSTED AS FOLLOWS:
506    * THE COEFFICIENTS 1.03, .718, .772 ARE THE SCRAP INPUT REQUIREMENTS IN THESE PROCESSES.  THIS
507    * IS NECESSARY AS THE SCRAP COLLECTION COSTS FOR THESE PROCESSES (PARAMETER SPR) ARE GIVEN IN
508    * TERMS OF THE PRICE OF COPPER.
509
510     OPM(IRUN,PM)$MAPIP(IRUN,PM)        = OPM(IRUN,PM) + 1.03*SPR(PM);
511     OPM(IRUN,PM)$(NOT MAPIP(IRUN,PM)) = 0;
512     OPM(IRUN,PM)                       = SUM(M, B(M,PM)*OPM(IRUN,PM));
513
```

```
GAMS 1.208  MODELING INVESTMENTS IN THE WORLD COPPER SECTOR                              02/16/84    14.28.26.    PAGE   14
            OPERATING COSTS


 514     OPS(JRUN,"TUBE-SCRAP")    = OPS(JRUN,"TUBE-REF-C") + SPR("SEMI")*.718/1000;
 515     OPS(JRUN,"S-SCRAP")       = OPS(JRUN,"S-REF-C")    + SPR("SEMI")*.772/1000;
 516     OPS(JRUN,P)               = SUM(M, B(M,P)*OPS(JRUN,P));
 517
 518     OPI(IRUN) = 1;
 519     OPJ(JRUN) = 1;
 520
 521     DISPLAY OPM,OPS;
```

```
GAMS 1.208  MODELING INVESTMENTS IN THE WORLD COPPER SECTOR                              02/16/84    14.28.26.    PAGE   15
            TRANSPORT


 523    * THE FOLLOWING PORTS WERE USED FOR COMPUTING DISTANCES:
 524    * AUSTRALIA                  : BUNBURY          C-AFRICA,ZAIRE,ZAMBIA    : BANANA
 525    * CANADA,EASTERN-US,USA      : NEW-YORK         CHILE,WS-AMERICA         : VALPARAISO
 526    * CHINA                      : SHANGHAI         EE+USSR                  : LENINGRAD
 527    * ES-AMERICA                 : RIO-DE-JANERIO   JAPAN+KOR                : TOKYO
 528    * MEX+CAM                    : VERACRUZ         O-ASIA                   : BOMBAY/CALCUTTA
 529    * PAPUA-NG                   : PORT MORESBY     PHILIPINES               : MANILA
 530    * PERU                       : CALLAO           S-AFRICA                 : RICHMUND BAY
 531    * W-EUROPE                   : ROTTERDAM        WESTERN-US               : PORTLAND
 532
 533     SET N OF NODES - ALL LOCATIONS;  N(IRUN) = YES; N(JRUN) = YES; ALIAS(N,NP,NPP);
 534
 535     TABLE UC(*,*)  TRANSPORT COST OF SEMI-MANUFACTURES AND ALL OTHER TYPES OF INTERMEDIATES
 536
 537                          OCEAN-FIX              OCEAN-VAR                     RAIL
 538    *                  (US$ PER TON)   (US$ PER TON PER NAUT. MI)  (US$ PER TON PER MILE)
 539
 540     ORE                        3                  .007                         .04
 541     B+R                        4                  .01                          .05
 542     SEMI                       8                  .02                          .1
 543
 544
 545    * THE DISTANCE MATRIX BELOW IS A LITTLE CONFUSING AS IT REPRESENTS DISTANCES BETWEEN ALL NODES
 546    * (UNION OF IRUN AND JRUN).
 547    *
 548     TABLE DISTANCE(*,*)   SEA DISTANCES IN NAUTICAL MILES
 549
 550                             AUSTRALIA   C-AFRICA     CANADA      CHILE      CHINA EASTERN-US    EE+USSR ES-AMERICA  JAPAN+KOR    MEX+CAM
 551
 552     C-AFRICA             9477
 553     CANADA              11409       5325
 554     CHILE                8358       7795       4634
 555     CHINA                3766      10205      10584      10148
 556     EASTERN-US          11409       5325                  4634      10584
 557     EE+USSR             10361       5897       4711       8692      12068       4711
 558     ES-AMERICA          10238       4125       4770       3670      11109       4770       6538
 559     JAPAN+KOR            4354      10974       9700       9280       1117       9700      12772      11513
 560     MEX+CAM             10854       6017       2023       4079       9463       2023       6522       4079       9155
 561     O-ASIA               4156       7103      11398      10275       4648      11398      12063       7848       4538      11927
 562     PAPUA-NG             1754       9184      10388       8400       2166      10388      11600       9945       2600       9100
 563     PERU                 9597       7096       3368       1306       8872       3368       7430       4976       8934       2813
 564     PHILIPINES           2600       9284      11388       9400       1166      11388      10952      10009       1770      10000
 565     S-AFRICA             6971       2506       6801       5694       7908       6801       7482       3267       8478       7346
 566     USA                 11409       5325                  4634      10584                  4711       4770       9700       2023
 567     W-EUROPE             9700       4657       3473       7454      10880       3473       2146       5300      11484       5284
 568     WESTERN-US           8742       9615       5887       5764       5440       5887       9949       8353       4323       5332
 569     WS-AMERICA           8358       7795       4634                 10148       4634       8692       3670       9280       4079
 570     ZAIRE                9477                  5325       7795      10205       5325       5897       4125      10974       6017
 571     ZAMBIA               9477                  5325       7795      10205       5325       5897       4125      10974       6017
```

```
GAMS 1.208  MODELING INVESTMENTS IN THE WORLD COPPER SECTOR                                    02/16/84     14.28.26.   PAGE   16
            TRANSPORT


  573     +               O-ASIA      PAPUA-NG     PERU     PHILIPINES    S-AFRICA      USA     W-EUROPE   WESTERN-US   WS-AMERICA
  574
  575     PAPUA-NG          5900
  576     PERU             11581         9100
  577     PHILIPINES        5900         1000       7706
  578     S-AFRICA          4581         6678       7000        6742
  579     USA              11398        10388       3368       11388        6801
  580     W-EUROPE         11066        10400       6192        9714        6244        3473
  581     WESTERN-US        7509         6923       4611        6024       10377        5887       8711
  582     WS-AMERICA       10275         8710       1306        9400        5694        4634       7454        5764
  583     ZAIRE             7104         9184       7096        9284        2506        5325       4657        9615         7795
  584     ZAMBIA            7104         9184       7096        9284        2506        5325       4657        9615         7795
  585
  586
  587     TABLE DRAIL   RAIL DISTANCES (KM)
  588
  589                         ZAMBIA      ZAIRE      EE+USSR    CANADA
  590
  591     MEX+CAM               1002        902
  592     S-AFRICA              1125        902
  593     CANADA                1002        902       1800
  594     WESTERN-US            1002        902
  595     EASTERN-US            1002        902                   350
  596     EE+USSR               1002        902
  597     AUSTRALIA             1250       1350       1800
  598     W-EUROPE              1002        902
  599     JAPAN+KOR             1250       1350       1800
  600     USA                   1002        902                   350
  601     ES-AMERICA            1002        902
  602     WS-AMERICA            1002        902
  603     C-AFRICA              1250       1350
  604     O-ASIA                1250       1350       1800
  605     CHINA                 1250       1350       1800
  606     PAPUA-NG              1250       1350
  607     PHILIPINES            1250       1350
  608                                                                   ;
  609
  610     DISTANCE(N,NP) = MAX(DISTANCE(N,NP) , DISTANCE(NP,N)); DRAIL(N,NP) = MAX(DRAIL(N,NP) , DRAIL(NP,N));
  611     DISPLAY DISTANCE,DRAIL;
  612
  613
  614    * THE DISTANCE MATRIX ABOVE HAS SOME ERRORS (INCONSISTENCIES IN DISTANCES) AS THE DISTANCES WERE COMPUTED BY
  615    * HAND.  FOR EXAMPLE IF THERE ARE 3 POINTS (A,B,C), IT WAS FOUND THAT THE DISTANCE BETWEEN A AND C IS MUCH
  616    * GREATER THAN BY GOING FROM A TO C THROUGH B.  THERE MIGHT ALSO BE NUMEROUS SUCH INTERMEDIATE POINTS B FOR
  617    * SOME (A,C) COMBINATION.  THUS WE DETERMINE THE LARGEST SUCH DISCREPANCY AND REDUCE THE DIRECT DISTANCE BETWEEN
  618    * A AND C BY THIS VALUE.  THIS METHOD ONLY REMOVES THE FIRST LEVEL ERRORS.  THE FOLLOWING COMPUTATION DOES THIS:
  619
  620
  621     DISTANCE(N,NP) = DISTANCE(N,NP) - SMAX(NPP, MAX(0, DISTANCE(N,NP) - DISTANCE(N,NPP) - DISTANCE(NPP,NP)) );
  622
  623
  624     PARAMETER MUR(IRUN,IRUN,CIM) TRANSPORT COST: RAW MATERIAL & INTERMEDIATE GOODS                      (US$ PER TON)
  625               MUI(IRUN,JRUN)     TRANSPORT COST: REFINED COPPER TO SEMI-MANUFACTURE AND MARKETS      (US$ PER TON)
  626               MUFS(JRUN,JRUN)    TRANSPORT COST: FINAL: SEMI-MANUFACTURES FROM LOCATIONS TO MARKETS (US$ PER TON);
  627
  628
```

```
GAMS 1.208  MODELING INVESTMENTS IN THE WORLD COPPER SECTOR                                    02/16/84     14.28.26.   PAGE   17
            TRANSPORT


  629     MUR(IRUN,IRUNP,"ORE")       = UC("ORE","OCEAN-FIX")$DISTANCE(IRUN,IRUNP)
  630                                 + UC("ORE","OCEAN-VAR")*DISTANCE(IRUN,IRUNP) + UC("ORE","RAIL")*DRAIL(IRUN,IRUNP);
  631
  632     MUR(IRUN,IRUNP,"BLISTER")   = UC("B+R","OCEAN-FIX")$DISTANCE(IRUN,IRUNP)
  633                                 + UC("B+R","OCEAN-VAR")*DISTANCE(IRUN,IRUNP) + UC("B+R","RAIL")*DRAIL(IRUN,IRUNP);
  634
  635     MUI(IRUN,JRUN)              = UC("B+R","OCEAN-FIX")$DISTANCE(IRUN,JRUN)
  636                                 + UC("B+R","OCEAN-VAR")*DISTANCE(IRUN,JRUN) + UC("B+R","RAIL")*DRAIL(IRUN,JRUN);
  637
  638     MUFS(JRUN,JRUNP)            = UC("SEMI","OCEAN-FIX")$DISTANCE(JRUN,JRUNP)
  639                                 + UC("SEMI","OCEAN-VAR")*DISTANCE(JRUN,JRUNP) + UC("SEMI","RAIL")*DRAIL(JRUN,JRUNP);
  640
  641     DISPLAY MUR,MUI,MUFS;
```

```
GAMS 1.208  MODELING INVESTMENTS IN THE WORLD COPPER SECTOR                                  02/16/84     14.28.26.    PAGE   18
            PRICES AND TARIFFS


 643    SET NTFR(I,J) NO TARIFFS ON REFINED COPPER BETWEEN / (PERU,CHILE).WS-AMERICA    , (ZAMBIA,ZAIRE).C-AFRICA
 644                                                         MEX+CAM.MEX+CAM             , S-AFRICA.S-AFRICA
 645                                                         (WESTERN-US,EASTERN-US).USA, CANADA.CANADA
 646                                                         EE+USSR.EE+USSR             , AUSTRALIA.AUSTRALIA
 647                                                         W-EUROPE.W-EUROPE           , JAPAN+KOR.JAPAN+KOR      /;
 648
 649
 650    PARAMETER PC            COMMODITY PRICES (US$ PER TON) / REFINED-CU   2000 , SEMI  2200  /
 651              TARIFFR(I,J) TARIFFS ON REFINED COPPER  (US$ PER TON OF COPPER CONTENT)
 652              TARIFFS(J,J) TARIFFS ON SEMI-MANUFACTURED GOODS (US$ PER TON  OF COPPER CONTENT);
 653
 654    TABLE ITP(J,*)  IMPORT TARIFFS (%)
 655
 656                    REFINED-CU     SEMI
 657
 658    USA                  1           6
 659    MEX+CAM             15          15
 660    ES-AMERICA          15          15
 661    WS-AMERICA          15          15
 662    W-EUROPE                         8
 663    EE+USSR             15          15
 664    C-AFRICA            15          15
 665    S-AFRICA            15          15
 666    O-ASIA              15          15
 667    JAPAN+KOR            8.5        15
 668    CHINA               15          15
 669    AUSTRALIA                       15
 670    CANADA                           6
 671                                     ;
 672    TARIFFS(JP,J)                 = PC("SEMI")*ITP(J,"SEMI")/100;
 673    TARIFFS(J,J)                  = 0;
 674
 675    TARIFFR(I,J)$(NOT NTFR(I,J)) = PC("REFINED-CU")*ITP(J,"REFINED-CU")/100;
 676    DISPLAY TARIFFS,TARIFFR;
```

```
GAMS 1.208  MODELING INVESTMENTS IN THE WORLD COPPER SECTOR                                  02/16/84     14.28.26.    PAGE   19
            MODEL DEFINITION


 678    VARIABLES ZM(P,I,T)       PROCESS LEVEL: MINES SMELTERS AND REFINERIES                               (1000 TPY)
 679              ZS(P,J,T)       PROCESS LEVEL: WIRE SHEET AND TUBE PLANTS                                  (1000 TPY)
 680              XI(C,I,I,T)     INTERPLANT SHIPMENTS OF ORE AND BLISTER                                    (1000 TPY)
 681              XIR(I,J,T)      SHIPMENTS: REFINED COPPER FROM SMELTERS TO SEMI-MANUFACTURERS              (1000 TPY)
 682              XFR(I,J,T)      SHIPMENTS: REFINED COPPER TO MARKETS FOR END USE                           (1000 TPY)
 683              XFS(C,J,J,T)    SHIPMENTS: SEMI-MANUFACTURES TO MARKETS                                    (1000 TPY)
 684
 685              SSR(C,I,T)      SCRAP SUPPLY: SMELTERS AND REFINERIES                                      (1000 TPY)
 686              SSM(C,J,T)      SCRAP SUPPLY: SHEET AND TUBE PLANTS                                        (1000 TPY)
 687              SSA(J,T)        SCRAP SUPPLY: SEMI-MANUFACTURING                                           (1000 TPY)
 688
 689              HM(M,I,T)       CAPACITY EXPANSION: MINES SMELTERS AND REFINERIES                          (1000 TPY)
 690              SM(M,I,T)       UNUSED ECONOMIES-OF-SCALE EXPANSION: MINES SMELTERS AND REFINERIES         (1000 TPY)
 691              HS(M,J,T)       CAPACITY EXPANSION: WIRE TUBE AND SHEET PLANTS                             (1000 TPY)
 692              SS(M,J,T)       UNUSED ECONOMIES-OF-SCALE EXPANSION: WIRE TUBE AND SHEET PLANTS            (1000 TPY)
 693              YM(M,I,T)       EXPANSION DECISION VARIABLE: MINES SMELTERS AND REFINERIES
 694              YS(M,J,T)       EXPANSION DECISION VARIABLE: WIRE TUBE AND SHEETS PLANTS
 695
 696              PHIKM(T)        COSTS: CAPITAL CHARGES AT MINES SMELTERS AND REFINERIES                 (MILLION US$)
 697              PHIKS(T)        COSTS: CAPITAL CHARGES AT WIRE TUBE AND SHEET PLANTS                    (MILLION US$)
 698              PHIOM(T)        COSTS: OPERATING COSTS AT MINES SMELTERS AND REFINERIES                 (MILLION US$)
 699              PHIOS(T)        COSTS: OPERATING COSTS AT WIRE TUBE AND SHEET PLANTS                    (MILLION US$)
 700              PHIT(T)         COSTS: TRANSPORT COSTS                                                  (MILLION US$)
 701              PHITF(T)        COSTS: TARIFF COSTS                                                     (MILLION US$)
 702              PHIU(T)         COSTS: TOTAL ANNUAL UNDISCOUNTED COST                                   (MILLION US$)
 703              PHIUTF(T)       COSTS: TOTAL ANNUAL UNDISCOUNTED COST WITH TARIFFS                      (MILLION US$)
 704              PHI1            TOTAL COST                                                              (MILLION US$)
 705              PHI2            TOTAL COST WITH TARIFFS                                                 (MILLION US$)
 706
 707    POSITIVE VARIABLE ZM,ZS,XI,XIR,XFR,XFS,SSR,SSM,SSA,HM,HS,SM,SS
 708    BINARY   VARIABLE YM,YS
 709
 710    EQUATIONS MBM(C,I,T)      MATERIAL BALANCE: MINES SMELTERS AND REFINERIES                            (1000 TPY)
 711              MBS(C,J,T)      MATERIAL BALANCE: SEMI-MANUFACTURING                                       (1000 TPY)
 712              MR(C,J,T)       MARKET REQUIREMENTS                                                        (1000 TPY)
 713              OREC(P,I)       HIGH-GRADE ORE MINING LIMITATIONS                                          (1000 TPY)
 714              SBS(J,T)        SCRAP BALANCE AT SEMI-MANUFACTURING LOCATIONS                              (1000 TPY)
 715
 716              CCM(M,I,T)      CAPACITY CONSTRAINT: MINES SMELTERS AND REFINERIES                         (1000 TPY)
 717              CCS(M,J,T)      CAPACITY CONSTRAINT: SEMI-MANUFACTURING                                    (1000 TPY)
 718              ICM1(M,I,T)     MAXIMUM EXPANSION: MINES SMELTERS AND REFINERIES                           (1000 TPY)
 719              ICM2(M,I,T)     LIMITS TO ECONOMIES-OF-SCALE: MINES SMELTERS AND REFINERIES                (1000 TPY)
 720              ICS1(M,J,T)     MAXIMUM EXPANSION: WIRE TUBE AND SHEET PLANTS                              (1000 TPY)
 721              ICS2(M,J,T)     LIMITS TO ECONOMIES-OF-SCALE: WIRE TUBE AND SHEET PLANTS                   (1000 TPY)
 722
```

```
723          AKM(T)         ACCOUNTING: CAPITAL CHARGES FOR MINES SMELTERS AND REFINERIES      (MILLION US$)
724          AKS(T)         ACCOUNTING: CAPITAL CHARGES FOR WIRE TUBE AND SHEET PLANTS         (MILLION US$)
725          AOM(T)         ACCOUNTING: OPERATING COST FOR MINES SMELTERS AND REFINERIES       (MILLION US$)
726          AOS(T)         ACCOUNTING: OPEARTING COST FOR WIRE TUBE AND SHEET PLANTS          (MILLION US$)
727          AOT(T)         ACCOUNTING: TRANSPORT                                              (MILLION US$)
728          AOTF(T)        ACCOUNTING: TARIFFS                                                (MILLION US$)
729          AU(T)          ACCOUNTING: UNDISCOUNTED ANNUAL COST                               (MILLION US$)
730          AUTF(T)        ACCOUNTING: UNDISCOUNTED ANNUAL COST WITH TARIFSF                  (MILLION US$)
731          AOBJ           OBJECTIVE FUNCTION                                                 (MILLION US$)
732          AOBJTF         OBJECTIVE FUNCTION WITH TARIFFS                                    (MILLION US$);
```



```
 735   MBM(CM,I,T)$MAPIC(I,CM)..

 736    SUM(PM$MAPIP(I,PM), A(CM,PM)*ZM(PM,I,T)) + SUM(IP, XI(CM,IP,I,T))$CIM(CM) + SSR(CM,I,T)$CIL(CM)

 737      =G= SUM(J, XFR(I,J,T))$CFR(CM) + SUM(IP, XI(CM,I,IP,T))$CIM(CM) + SUM(J, XIR(I,J,T))$CFR(CM);

 738

 739   MBS(CS,J,T)$MAPJC(J,CS)..   SUM(PSM$MAPJP(J,PSM), A(CS,PSM)*ZS(PSM,J,T)) + SUM(I, XIR(I,J,T))$CFR(CS)

 740                             + SSM(CS,J,T)$CIL(CS) =G= SUM(JP, XFS(CS,J,JP,T))$CFS(CS) ;

 741

 742   MR(CF,J,T)..  SUM(I, XFR(I,J,T))$CFR(CF) + SUM(JP, XFS(CF,JP,J,T))$CFS(CF) =G= DEMAND(J,CF,T);

 743

 744   CCM(MM,I,T)..  SUM(PM$MAPIP(I,PM), B(MM,PM)*ZM(PM,I,T)) =L= CAPM(I,MM) + SUM(TP$TS(T,TP), HM(MM,I,TP));

 745

 746   CCS(MS,J,T)..  SUM(PSM$MAPJP(J,PSM), B(MS,PSM)*ZS(PSM,J,T)) =L= CAPS(J,MS) + SUM(TP$TS(T,TP), HS(MS,J,TP));

 747

 748   ICM1(MM,I,T)..  HM(MM,I,T) =L= HHATM(I,MM)*YM(MM,I,T);

 749

 750   ICS1(MS,J,T)..  HS(MS,J,T) =L= HHATS(J,MS)*YS(MS,J,T);

 751

 752   ICM2(MM,I,T)..  HM(MM,I,T) + SM(MM,I,T) =G= HBARM(I,MM)*YM(MM,I,T);

 753

 754   ICS2(MS,J,T)..  HS(MS,J,T) + SS(MS,J,T) =G= HBARS(J,MS)*YS(MS,J,T);

 755

 756   OREC(PMM,I)..  SUM(T, RPH(T)*ZM(PMM,I,T)) =L= 1000*RESERVES(I,PMM);

 757

 758   SBS(J,T).. SSA(J,T) =E= SUM(CIL, SSM(CIL,J,T));

 759

 760   AKM(T).. PHIKM(T) =E= SIGMA*SUM(TP$TS(T,TP), SUM((I,MM), SFI(I)*(OMEGAM(I,MM)*SM(MM,I,TP) + NUM(I,MM)*HM(MM,I,TP))));

 761

 762   AKS(T).. PHIKS(T) =E= SIGMA*SUM(TP$TS(T,TP), SUM((J,MS), SFJ(J)*(OMEGAS(J,MS)*SS(MS,J,TP) + NUS(J,MS)*HS(MS,J,TP))));
```

```
GAMS 1.208  MODELING INVESTMENTS IN THE WORLD COPPER SECTOR                              02/16/84     14.28.26.    PAGE   21
            MODEL DEFINITION


 763

 764   AOM(T).. PHIOM(T) =E= SUM((PM,I)$MAPIP(I,PM), OPI(I)*OPM(I,PM)*ZM(PM,I,T)) / 1000;

 765

 766   AOS(T).. PHIOS(T) =E= SUM((PSM,J), OPJ(J)*OPS(J,PSM)*ZS(PSM,J,T));

 767

 768   AOT(T).. PHIT(T) =E= ( SUM((CIM,I,IP), MUR(I,IP,CIM)*XI(CIM,I,IP,T)) + SUM((CFS,J,JP), MUFS(J,JP)*XFS(CFS,J,JP,T))

 769                        + SUM((I,J), MUI(I,J)*(XIR(I,J,T) + XFR(I,J,T))) ) / 1000;

 770

 771   AOTF(T).. PHITF(T) =E= SUM(J, SUM(I, TARIFFR(I,J)*(XFR(I,J,T) + XIR(I,J,T)))

 772                               + SUM((CFS,JP), TARIFFS(JP,J)*XFS(CFS,JP,J,T))  )/1000;

 773

 774   AU(T)..  PHIU(T) =E= PHIKM(T) + PHIKS(T) + PHIOM(T) + PHIOS(T) + PHIT(T);

 775

 776   AUTF(T)..  PHIUTF(T) =E= PHIKM(T) + PHIKS(T) + PHIOM(T) + PHIOS(T) + PHIT(T) + PHITF(T);

 777

 778   AOBJ.. PHI1 =E= SUM(T, PERIOD(T)*DIS(T)*PHIU(T));

 779

 780   AOBJTF.. PHI2 =E= SUM(T, PERIOD(T)*DIS(T)*PHIUTF(T));

 781
 783  * BOUNDS
 784   ZM.UP(PMH,I,T)     = RESC(I,PMH);
 785   ZS.LO("WIRE-REF-C",J,T) = CAPS(J,"WIRE");
 786
 787   SSR.UP(CIL,I,T) = SCRAPI(I,T,CIL);
 788
 789  * DEPENDING ON SEMI-MANUFACTURING SCRAP FORMULATION USE EITHER:
 790  *SSM.UP(CIL,J,T)    = SCRAPJ(J,T,CIL);
 791  * OR:
 792   SSA.UP(J,T) = SCRAPJ(J,T,"SCRAP");
 793
 794   MODEL COPPER / ALL /;
 795
 796   SOLVE COPPER MINIMIZING PHI2 USING MIP;



GAMS 1.208  MODELING INVESTMENTS IN THE WORLD COPPER SECTOR                              02/16/84     14.28.26.    PAGE   22
            REPORT

 798   SET CPS  / EXIST-CAP, NEW-CAP, TOTAL-CAP, PRODUCTION, SHIPPED, SLACK /
 799
 800   PARAMETER XII(*,*,*)    SHIPMENTS OF INPUTS (ORE-BLISTER-REFINED COPPER - 1000 TONS)
 801             XFD(*,*,C)    SHIPMENTS OF FINAL PRODUCTS: DISAGGREGATED       (1000 TONS)
 802
 803             CPDM          CAPACITY-PRODUCTION AT MINES          (1000 TONS)
 804             CPDS          CAPACITY-PRODUCTION AT SMELTERS       (1000 TONS)
 805             CPDR          CAPACTIY-PRODUCTION AT REFINERIES     (1000 TONS)
 806             CPDW          CAPACITY-PRODUCTION AT WIRE PLANTS    (1000 TONS)
 807             CPDT          CAPACITY-PRODUCTION AT TUBE PLANTS    (1000 TONS)
 808             CPDSH         CAPACITY-PRODUCTION AT SHEET PLANTS   (1000 TONS);
 809
 810   XII(I,IP,CIM)           = SUM(T, XI.L(CIM,I,IP,T));
 811   XII(I,I,"ORE")          = SUM(T, SUM(PMM, ZM.L(PMM,I,T)) - SUM(IP, XI.L("ORE",I,IP,T)) );
 812   XII(I,I,"BLISTER")      = SUM(T, SUM(PS, ZM.L(PS,I,T)) - SUM(IP, XI.L("BLISTER",I,IP,T)) );
 813   XII(I,J,"REFINED-CU")   = SUM(T, XIR.L(I,J,T));
 814   XII(I,"TOTAL",CIM)      = SUM(IP, XII(I,IP,CIM));
 815   XII(I,"TOTAL",CFR)      = SUM(J, XII(I,J,CFR));
 816   XII("TOTAL",I,CIM)      = SUM(IP, XII(IP,I,CIM));
 817   XII("TOTAL",J,CFR)      = SUM(I, XII(I,J,CFR));
 818   XII("TOTAL","TOTAL",C) = SUM(I, XII(I,"TOTAL",C));
 819
 820   XFD(I,J,"REFINED-CU")   = SUM(T, XFR.L(I,J,T));
 821   XFD(J,JP,CFS)           = SUM(T, XFS.L(CFS,J,JP,T));
 822   XFD(I,"TOTAL",CFR)      = SUM(J, XFD(I,J,CFR));
 823   XFD("TOTAL",J,CFR)      = SUM(I, XFD(I,J,CFR));
 824   XFD(J,"TOTAL",CFS)      = SUM(JP, XFD(J,JP,CFS));
 825   XFD("TOTAL",J,CFS)      = SUM(JP, XFD(JP,J,CFS));
 826   XFD("TOTAL","TOTAL",C) = SUM(J, XFD("TOTAL",J,C));
 827   DISPLAY XII,XFD;
 828
 829
```

```
830    CPDM(I,"EXIST-CAP")      = CAPM(I,"OPEN-PIT");
831    CPDM(I,"NEW-CAP")        = SUM(T, HM.L("OPEN-PIT",I,T));
832    CPDM(I,"TOTAL-CAP")      = CPDM(I,"EXIST-CAP") + CPDM(I,"NEW-CAP");
833    CPDM(I,"PRODUCTION")     = SUM((T,PMM), ZM.L(PMM,I,T));
834    CPDM(I,"SHIPPED")        = XII(I,"TOTAL","ORE");
835    CPDM(I,"SLACK")          = ROUND(CPDM(I,"TOTAL-CAP") - SUM((T,PMM), B("OPEN-PIT",PMM)*ZM.L(PMM,I,T)),1);
836    CPDM(I,"CAP-UT")$CPDM(I,"TOTAL-CAP") = ROUND(100*SUM((T,PMM), B("OPEN-PIT",PMM)*ZM.L(PMM,I,T))/CPDM(I,"TOTAL-CAP"),1);
837    CPDM("**TOTAL**",CPS)    = SUM(I, CPDM(I,CPS));
838
839    CPDS(I,"EXIST-CAP")      = CAPM(I,"SMELTER");
840    CPDS(I,"NEW-CAP")        = SUM(T, HM.L("SMELTER",I,T));
841    CPDS(I,"TOTAL-CAP")      = CPDS(I,"EXIST-CAP") + CPDS(I,"NEW-CAP");
842    CPDS(I,"PRODUCTION")     = SUM((T,PS), ZM.L(PS,I,T));
843    CPDS(I,"SHIPPED")        = XII(I,"TOTAL","BLISTER");
844    CPDS(I,"SLACK")          = ROUND(CPDS(I,"TOTAL-CAP") - SUM((T,PS), B("SMELTER",PS)*ZM.L(PS,I,T)),1);
845    CPDS(I,"CAP-UT")$CPDS(I,"TOTAL-CAP") = ROUND(100*SUM((T,PS), B("SMELTER",PS)*ZM.L(PS,I,T))/CPDS(I,"TOTAL-CAP"),1);
846    CPDS("**TOTAL**",CPS)    = SUM(I, CPDS(I,CPS));
847
848    CPDR(I,"EXIST-CAP")      = CAPM(I,"REFINERY");
849    CPDR(I,"NEW-CAP")        = SUM(T, HM.L("REFINERY",I,T));
850    CPDR(I,"TOTAL-CAP")      = CPDR(I,"EXIST-CAP") + CPDR(I,"NEW-CAP");
851    CPDR(I,"PRODUCTION")     = SUM((T,PR), ZM.L(PR,I,T));
852    CPDR(I,"SHIPPED")        = XII(I,"TOTAL","REFINED-CU") + XFD(I,"TOTAL","REFINED-CU");
853    CPDR(I,"SLACK")          = ROUND(CPDR(I,"TOTAL-CAP") - SUM((T,PR), B("REFINERY",PR)*ZM.L(PR,I,T)),1);
```



```
854    CPDR(I,"CAP-UT")$CPDR(I,"TOTAL-CAP") = ROUND(100*SUM((T,PR), B("REFINERY",PR)*ZM.L(PR,I,T))/CPDR(I,"TOTAL-CAP"),1);
855    CPDR("**TOTAL**",CPS)    = SUM(I, CPDR(I,CPS));
856    DISPLAY CPDM,CPDS,CPDR;
857
858    CPDW(J,"EXIST-CAP")       = CAPS(J,"WIRE");
859    CPDW(J,"NEW-CAP")         = SUM(T, HS.L("WIRE",J,T));
860    CPDW(J,"TOTAL-CAP")       = CPDW(J,"EXIST-CAP") + CPDW(J,"NEW-CAP");
861    CPDW(J,"PRODUCTION")      = SUM((T,PW), ZS.L(PW,J,T));
862    CPDW(J,"SHIPPED")         = XFD(J,"TOTAL","WIRE");
863    CPDW(J,"SLACK")           = ROUND(CPDW(J,"PRODUCTION") - SUM((T,PW), B("WIRE",PW)*ZS.L(PW,J,T)),1);
864    CPDW(J,"CAP-UT")$CPDW(J,"TOTAL-CAP") = ROUND(100*SUM((T,PW), B("WIRE",PW)*ZS.L(PW,J,T))/CPDW(J,"TOTAL-CAP"),1);
865    CPDW("**TOTAL**",CPS)    = SUM(J, CPDW(J,CPS));
866    DISPLAY CPDW;
867
868    CPDT(J,"EXIST-CAP")      = CAPS(J,"TUBES+RODS");
869    CPDT(J,"NEW-CAP")        = SUM(T, HS.L("TUBES+RODS",J,T));
870    CPDT(J,"TOTAL-CAP")      = CPDT(J,"EXIST-CAP") + CPDT(J,"NEW-CAP");
871    CPDT(J,"PRODUCTION")     = SUM((T,PTR), ZS.L(PTR,J,T));
872    CPDT(J,"SHIPPED")        = XFD(J,"TOTAL","TUBES+RODS");
873    CPDT(J,"SLACK")          = ROUND(CPDT(J,"PRODUCTION") - SUM((T,PTR), B("TUBES+RODS",PTR)*ZS.L(PTR,J,T)),1);
874    CPDT(J,"CAP-UT")$CPDT(J,"TOTAL-CAP") = ROUND(100*SUM((T,PTR), B("TUBES+RODS",PTR)*ZS.L(PTR,J,T))/CPDT(J,"TOTAL-CAP")
875                                           ,1);
876    CPDT("**TOTAL**",CPS)    = SUM(J, CPDT(J,CPS));
877    DISPLAY CPDT;
878
879    CPDSH(J,"EXIST-CAP")     = CAPS(J,"SHEETS+P+S");
880    CPDSH(J,"NEW-CAP")       = SUM(T, HS.L("SHEETS+P+S",J,T));
881    CPDSH(J,"TOTAL-CAP")     = CPDSH(J,"EXIST-CAP") + CPDSH(J,"NEW-CAP") ;
882    CPDSH(J,"PRODUCTION")    = SUM((T,PSH), ZS.L(PSH,J,T));
883    CPDSH(J,"SHIPPED")       = XFD(J,"TOTAL","SHEETS+P+S");
884    CPDSH(J,"SLACK")         = ROUND(CPDSH(J,"PRODUCTION") - SUM((T,PSH), B("SHEETS+P+S",PSH)*ZS.L(PSH,J,T)),1);
885    CPDSH(J,"CAP-UT")$CPDSH(J,"TOTAL-CAP")
886                             = ROUND(100*SUM((T,PSH), B("SHEETS+P+S",PSH)*ZS.L(PSH,J,T))/CPDSH(J,"TOTAL-CAP"),1);
887    CPDSH("**TOTAL**",CPS) = SUM(J, CPDSH(J,CPS));
888    DISPLAY CPDSH;
889
890    PARAMETER SPRICE(J,CF,*)  SHADOW PRICE OF DEMAND BALANCE
891              COSTT(*)        COST TABULATION (MILLION US$);
892
893    SPRICE(J,CF,"$-PER-TON")  = - 1000*SUM(T, MR.M(CF,J,T));
894    SPRICE(J,CF,"$-PER-LB")   = SPRICE(J,CF,"$-PER-TON")/LBPERTON;
895    DISPLAY SPRICE;
896
897    COSTT("CAPITAL-I")       = SUM(T, PHIKM.L(T));
898    COSTT("CAPITAL-J")       = SUM(T, PHIKS.L(T));
899    COSTT("OPERATE-I")       = SUM(T, PHIOM.L(T));
900    COSTT("OPERATE-J")       = SUM(T, PHIOS.L(T));
901    COSTT("TRANSPORT")       = SUM(T, PHIT.L(T));
902    COSTT("TARIFF")          = SUM(T, PHITF.L(T));
903    COSTT("TOTAL-COST")      = PHI2.L;
904    DISPLAY COSTT;
```

```
GAMS 1.208  MODELING INVESTMENTS IN THE WORLD COPPER SECTOR              02/16/84    14.28.26.    PAGE   24
            REFERENCE MAP OF VARIABLES

VARIABLES   TYPE    REFERENCES

A           PARAM       REF      270      271      281      286      736      739  DEFINED      175      DCL
                        175
AKM         EQU     DEFINED      760      DCL      723
AKS         EQU     DEFINED      762      DCL      724
AOBJ        EQU     DEFINED      778      DCL      731
AOBJTF      EQU     DEFINED      780      DCL      732
AOM         EQU     DEFINED      764      DCL      725
AOS         EQU     DEFINED      766      DCL      726
AOT         EQU     DEFINED      768      DCL      727
AOTF        EQU     DEFINED      771      DCL      728
AU          EQU     DEFINED      774      DCL      729
AUTF        EQU     DEFINED      776      DCL      730
B           PARAM       REF      512      516      744      746      835      836      844      845      853
                        854      863      864      873      874      884      886  DEFINED      196      DCL
                        196
C           SET         REF       51       53       55       57       59       61       63      175      224
                        263      265      680      683      685      686      710      711      712      801
                        818      826  DEFINED       38  CONTROL      818      826      DCL       38
CAPM        PARAM       REF      744      830      839      848  DEFINED      335      DCL      335
CAPS        PARAM       REF      746      785      858      868      879  DEFINED      317      DCL      317
CCM         EQU     DEFINED      744      DCL      716
CCS         EQU     DEFINED      746      DCL      717
CF          SET         REF      146      170      171    4*742      890      893      894  DEFINED       55
                    CONTROL      170      171      742      893      894      DCL       55
CFR         SET         REF    2*737      739      742      815      817      822      823  DEFINED       57
                    CONTROL      815      817      822      823      DCL       57
CFS         SET         REF      740      742      768      772      821      824      825  DEFINED       59
                    CONTROL      768      772      821      824      825      DCL       59
CGR         SET         REF      144    2*170    2*171  DEFINED      141  CONTROL      170      171      DCL
                        141
CIL         SET         REF      254      255      256      270      271      279    2*281      284    2*286
                        736      740      758      787  DEFINED       63  CONTROL      254      255      256
                        270      271      279      281      284      286      758      787      DCL       63
CIM         SET         REF      624      736      737    2*768      810      814      816  DEFINED       61
                    CONTROL      768      810      814      816      DCL       61
CM          SET         REF      735    5*736    4*737  DEFINED       51  CONTROL      278      735      DCL
                         51
COPPER      MODEL       REF      796  DEFINED      794      DCL      794
COSTT       PARAM       REF      904  DEFINED      897      898      899      900      901      902      903
                        DCL      891
CPDM        PARAM       REF    2*832      835    2*836      837      856  DEFINED      830      831      832
                        833      834      835      836      837      DCL      803
CPDR        PARAM       REF    2*850      853    2*854      855      856  DEFINED      848      849      850
                        851      852      853      854      855      DCL      805
CPDS        PARAM       REF    2*841      844    2*845      846      856  DEFINED      839      840      841
                        842      843      844      845      846      DCL      804
CPDSH       PARAM       REF    2*881      884      885      886      887      888  DEFINED      879      880
                        881      882      883      884      885      887      DCL      808
CPDT        PARAM       REF    2*870      873    2*874      876      877  DEFINED      868      869      870
                        871      872      873      874      876      DCL      807
CPDW        PARAM       REF    2*860      863    2*864      865      866  DEFINED      858      859      860
                        861      862      863      864      865      DCL      806
CPS         SET         REF      837      846      855      865      876      887  DEFINED      798  CONTROL
                        837      846      855      865      876      887      DCL      798
```

```
GAMS 1.208  MODELING INVESTMENTS IN THE WORLD COPPER SECTOR              02/16/84    14.28.26.    PAGE   25
            REFERENCE MAP OF VARIABLES

VARIABLES   TYPE    REFERENCES

CS          SET         REF    3*739    4*740  DEFINED       53  CONTROL      283      739      DCL       53
DEMAND      PARAM       REF      171      173      742  DEFINED      170      171      DCL      146
DEM1980     PARAM       REF    2*170      171  DEFINED      149      DCL      149
DIS         PARAM       REF      456      778      780  DEFINED      451      452      DCL      379
DISTANCE    PARAM       REF    2*610      611    4*621      629      630      632      633      635      636
                        638      639  DEFINED      548      610      621      DCL      548
DRAIL       PARAM       REF    2*610      611      630      633      636      639  DEFINED      587      610
                        DCL      587
HBARM       PARAM       REF      437    2*440    2*441      456      752  DEFINED      433      434      DCL
                        387
HBARS       PARAM       REF      438    2*442      456      754  DEFINED      435      DCL      388
HDS1        PARAM       REF    2*254      279  DEFINED      228      DCL      228
HDS2        PARAM       REF    2*255      284  DEFINED      242      DCL      242
HHATM       PARAM       REF      748  DEFINED      437      DCL      389
HHATS       PARAM       REF      750  DEFINED      438      DCL      390
HM          VAR         REF      707      744      748      752      760      831      840      849      DCL
                        689
HS          VAR         REF      707      746      750      754      762      859      869      880      DCL
                        691
I           SET         REF      118      643      651      675      678    2*680      681      682      685
                        689      690      693      710      713      716      718      719      735    4*736
                      3*737      739      742    4*744    3*748    4*752    2*756    5*760    4*764    2*768
                      3*769    3*771      784      787      810    3*811    3*812      813      814      815
                        816      817      818      820      822      823      830      831    2*832      833
```

```
                         834     2*835     3*836       837       839       840     2*841       842       843     2*844
                       3*845       846       848       849     2*850       851     2*852     2*853     3*854       855
                     DEFINED        25   CONTROL       675       735       739       742       744       748       752
                         756       760       764       768       769       771       784       787       810       811
                         812       813       814       815       816       817       818       820       822       823
                         830       831       832       833       834       835       836       837       839       840
                         841       842       843       844       845       846       848       849       850       851
                         852       853       854       855       DCL        25
ICM1        EQU      DEFINED       748       DCL       718
ICM2        EQU      DEFINED       752       DCL       719
ICS1        EQU      DEFINED       750       DCL       720
ICS2        EQU      DEFINED       754       DCL       721
IH          SET          REF       367   DEFINED       355   CONTROL       370       DCL       355
IIT         SET          REF       373       433       440       444   DEFINED     2*370       DCL       363
IL          SET          REF       373   DEFINED       367   CONTROL       370       DCL       357
INVM        PARAM        REF       434       441       445   DEFINED       397       DCL       397
INVRS       PARAM        REF       433       440       444   DEFINED       421       DCL       421
INVS        PARAM        REF       435       442       446   DEFINED       408       DCL       408
IP          SET          REF       736       737     2*768       810       811       812       814       816   CONTROL
                         736       737       768       810       811       812       814       816       DCL       118
IRUN        SET          REF        25       118       224       228     2*254       263       264       279       281
                         290       292       313       335       355       357       363       367       393       397
                         433       434       437     3*440     3*441     2*444     2*445       458       464       475
                         480       485     2*501     2*502     2*503     2*510       511       512     2*624       625
                         629     2*630       632     2*633       635     2*636   DEFINED        20   CONTROL       254
                         278       279       280       281       313       367       433       434       437       440
                         441       444       445       448       501       502       503       510       511       512
                         518       533       629       632       635       DCL        20
IRUNP       SET          REF       629     2*630       632     2*633   CONTROL       629       632       DCL       118
IT          SET          REF       363       421     2*433     2*440     2*444   DEFINED       362   CONTROL       433
```



```
VARIABLES   TYPE     REFERENCES

                         440       444       DCL       362
ITP         PARAM        REF       672       675   DEFINED       654       DCL       654
J           SET          REF       118       643       651     2*652       654       672       673     2*675       679
                         681       682     2*683       686       687       691       692       694       711       712
                         714       717       720       721     2*737     4*739     2*740     3*742     4*746     3*750
                       4*754     2*758     5*762     3*766     2*768     3*769     3*771     2*772       785       792
                         813       815       817       820       821       822       823       824       825       826
                         858       859     2*860       861       862     2*863     3*864       865       868       869
                       2*870       871       872     2*873     3*874       876       879       880     2*881       882
                         883     2*884       885     2*886       887       890       893       894   DEFINED        34
                     CONTROL       672       673       675     2*737       739       742       746       750       754
                         758       762       766       768       769       771       785       792       813       815
                         817       820       821       822       823       824       825       826       858       859
                         860       861       862       863       864       865       868       869       870       871
                         872       873       874       876       879       880       881       882       883       884
                         885       887       893       894       DCL        34
JH          SET          REF       368   DEFINED       358   CONTROL       371       DCL       358
JL          SET          REF       373   DEFINED       368   CONTROL       371       DCL       360
JP          SET          REF       740       742     2*768     2*772       821       824       825   CONTROL       672
                         740       742       768       772       821       824       825       DCL       118
JRUN        SET          REF        34       118       146       149     2*170     2*171       225       242     2*255
                         256       265       266       284       286       317       358       360       365       368
                         394       435       438     3*442     2*446       487       490       514       515       516
                         625     2*626       635     2*636       638     2*639   DEFINED        30   CONTROL       170
                         171       255       256       283       284       285       286       368       435       438
                         442       446       449       514       515       516       519       533       635       638
                         DCL        30
JRUNP       SET          REF       638     2*639   CONTROL       638       DCL       118
JT          SET          REF       365       408     2*435     2*442     2*446   DEFINED       364   CONTROL       435
                         442       446       DCL       364
JTJ         SET          REF       373       435       442       446   DEFINED     2*371       DCL       365
LBPERTON    PARAM        REF       894   DEFINED       139       DCL       139
LIFE        PARAM        REF     2*453   DEFINED       377       DCL       377
M           SET          REF       107       109       119       196       317       335       408       421       433
                         435       437       438     3*440     3*442     2*444     2*446       512       516       689
                         690       691       692       693       694       716       717       718       719       720
                         721   DEFINED        98   CONTROL       119       433       435       437       438       440
                         442       444       446       512       516       DCL        98
MAPIC       PARAM        REF       281       288       735   DEFINED       278       279       DCL       263
MAPIP       PARAM        REF       288       501       502       503       510       511       736       744       764
                     DEFINED       280       281       DCL       264
MAPJC       PARAM        REF       286       288       739   DEFINED       283       284       DCL       265
MAPJP       PARAM        REF       288       739       746   DEFINED       285       286       DCL       266
MAX         FUNCT        REF     2*610       621
MBM         EQU      DEFINED       735       DCL       710
MBS         EQU      DEFINED       739       DCL       711
MGR         SET          REF       170       171   DEFINED       144       DCL       144
MM          SET          REF       119     3*744     3*748     4*752     4*760   DEFINED       107   CONTROL       744
                         748       752       760       DCL       107
MR          EQU          REF       893   DEFINED       742       DCL       712
MS          SET          REF     3*746     3*750     4*754     4*762   DEFINED       119   CONTROL       746       750
                         754       762       DCL       109
MUFS        PARAM        REF       641       768   DEFINED       638       DCL       626
MUI         PARAM        REF       641       769   DEFINED       635       DCL       625
```

```
GAMS 1.208  MODELING INVESTMENTS IN THE WORLD COPPER SECTOR                          02/16/84    14.28.26.    PAGE   27
            REFERENCE MAP OF VARIABLES

VARIABLES   TYPE     REFERENCES

MUR         PARAM         REF      641       768   DEFINED       629       632       DCL       624
N           SET           REF      533     4*610     3*621   DEFINED     2*533   CONTROL     2*610       621       DCL
                          533
NP          SET           REF    4*610     3*621   CONTROL     2*610       621       DCL       533
NPP         SET           REF    2*621   CONTROL       621       DCL       533
NTFR        SET           REF      675   DEFINED       643       DCL       643
NUM         PARAM         REF      456       760   DEFINED       444       445       DCL       384
NUS         PARAM         REF      456       762   DEFINED       446       DCL       385
OM          PARAM         REF      501   DEFINED       464       DCL       464
OMEGAM      PARAM         REF      444       445       456       760   DEFINED       440       441       DCL       382
OMEGAS      PARAM         REF      446       456       762   DEFINED       442       DCL       383
OPI         PARAM         REF      764   DEFINED       518       DCL       485
OPJ         PARAM         REF      766   DEFINED       519       DCL       487
OPM         PARAM         REF      510       512       521       764   DEFINED       501       502       503       510
                          511      512       DCL       458
OPR         PARAM         REF      503   DEFINED       480       DCL       480
OPS         PARAM         REF      514       515       516       521       766   DEFINED       490       514       515
                          516      DCL       490
OREC        EQU       DEFINED      756       DCL       713
OS          PARAM         REF      502   DEFINED       475       DCL       475
P           SET           REF       79        82        84        86        88        90        92        94        96
                          119      175       196       260       261       264       266       290       292       458
                          490    2*516       678       679       713   DEFINED        65   CONTROL       119       516
                          DCL       65
PC          PARAM         REF      672       675   DEFINED       650       DCL       650
PERIOD      PARAM         REF      131       133       778       780   DEFINED       130       137       DCL       126
PHIKM       VAR           REF      760       774       776       897       DCL       696
PHIKS       VAR           REF      762       774       776       898       DCL       697
PHIOM       VAR           REF      764       774       776       899       DCL       698
PHIOS       VAR           REF      766       774       776       900       DCL       699
PHIT        VAR           REF      768       774       776       901       DCL       700
PHITF       VAR           REF      771       776       902       DCL       701
PHIU        VAR           REF      774       778       DCL       702
PHIUTF      VAR           REF      776       780       DCL       703
PHI1        VAR           REF      778       DCL       704
PHI2        VAR           REF      780       796       903       DCL       705
PM          SET           REF      119       121       270     3*510       511     2*512     3*736     3*744     3*764
                      DEFINED       79   CONTROL       270       280       510       511       512       736       744
                          764      DCL        79
PMH         SET           REF      313       784   DEFINED        82   CONTROL       313       784       DCL        82
PMM         SET           REF      501     2*756       811       833     2*835     2*836   DEFINED        84   CONTROL
                          501      756       811       833       835       836       DCL        84
PR          SET           REF      503       851     2*853     2*854   DEFINED        88   CONTROL       503       851
                          853      854       DCL        88
PS          SET           REF      502       812       842     2*844     2*845   DEFINED        86   CONTROL       502
                          812      842       844       845       DCL        86
PSH         SET           REF      882     2*884     2*886   DEFINED        96   CONTROL       882       884       886
                          DCL       96
PSI         SET           REF      281   DEFINED       270   CONTROL       281       DCL       260
PSJ         SET           REF      286   DEFINED       271   CONTROL       286       DCL       261
PSM         SET           REF      121       271     3*739     3*746     2*766   DEFINED       119   CONTROL       271
                          285      739       746       766       DCL        90
PTR         SET           REF      871     2*873     2*874   DEFINED        94   CONTROL       871       873       874
                          DCL       94
```

```
GAMS 1.208  MODELING INVESTMENTS IN THE WORLD COPPER SECTOR                          02/16/84    14.28.26.    PAGE   28
            REFERENCE MAP OF VARIABLES

VARIABLES   TYPE     REFERENCES

PW          SET           REF      861     2*863     2*864   DEFINED        92   CONTROL       861       863       864
                          DCL       92
RESC        PARAM         REF      314       784   DEFINED       313       DCL       290
RESERVES    PARAM         REF      313       756   DEFINED       292       DCL       292
RHO         PARAM         REF      451     3*453   DEFINED       376       DCL       376
ROUND       FUNCT         REF      835       836       844       845       853       854       863       864       873
                          875      884       886
RPH         PARAM         REF      133       756   DEFINED       131       DCL       127
SBS         EQU       DEFINED      758       DCL       714
SCRAPI      PARAM         REF      258       787   DEFINED       254       DCL       224
SCRAPJ      PARAM         REF      256       258       792   DEFINED       255       256       DCL       225
SFI         PARAM         REF      760   DEFINED       448       DCL       393
SFJ         PARAM         REF      762   DEFINED       449       DCL       394
SIGMA       PARAM         REF      456       760       762   DEFINED       453       DCL       380
SM          VAR           REF      707       752       760       DCL       690
SPR         PARAM         REF      510       514       515   DEFINED       460       DCL       460
SPRICE      PARAM         REF      894       895   DEFINED       893       894       DCL       890
SS          VAR           REF      707       754       762       DCL       692
SSA         VAR           REF      707       758   DEFINED       792       DCL       687
SSM         VAR           REF      707       740       758       DCL       686
SSR         VAR           REF      707       736   DEFINED       787       DCL       685
T           SET           REF      118     2*130       131       392       451       455       678       679       680
                          681      682       683       685       686       687       689       690       691       692
                          693      694       696       697       698       699       700       701       702       703
```

```
                              710      711      712      714      716      717      718      719      720      721
                              723      724      725      726      727      728      729      730    3*736    3*737
                            2*739    2*740    3*742    2*744    2*746    2*748    2*750    3*752    3*754    2*756
                            2*758    2*760    2*762    2*764    2*766    3*768    2*769    3*771      772    6*774
                            7*776    3*778    3*780      787      792      810    2*811    2*812      813      820
                              821      831      833      835      836      840      842      844      845      849
                              851      853      854      859      861      863      864      869      871      873
                              874      880      882      884      886      893      897      898      899      900
                              901      902  DEFINED      113  CONTROL      130      131      137      451      452
                              455      735      739      742      744      746      748      750      752      754
                              756      758      760      762      764      766      768      771      774      776
                              778      780      784      785      787      792      810      811      812      813
                              820      821      831      833      835      836      840      842      844      845
                              849      851      853      854      859      861      863      864      869      871
                              873      874      880      882      884      886      893      897      898      899
                              900      901      902      DCL      113
TARIFFR     PARAM             REF      676      771  DEFINED      675      DCL      651
TARIFFS     PARAM             REF      676      772  DEFINED      672      673      DCL      652
TBASE       SET               REF      129  DEFINED      115      DCL      115
TG          SET               REF      113      115      118      125      126      127      129      142      143
                              144      146      224      225      254      255      256  DEFINED      111  CONTROL
                              129      254      255      256      DCL      111
TGP         SET               REF    2*129  CONTROL      129      DCL      118
TG1         SET               REF    2*170  DEFINED      142  CONTROL      170      DCL      142
TG2         SET               REF    2*171  DEFINED      143  CONTROL      171      DCL      143
TP          SET               REF      392      455    2*744    2*746    3*760    3*762  CONTROL      455      744
                              746      760      762      DCL      118
TS          PARAM             REF      456      744      746      760      762  DEFINED      455      DCL      392
UC          PARAM             REF      629    2*630      632    2*633      635    2*636      638    2*639  DEFINED
                              535      DCL      535


GAMS 1.208  MODELING INVESTMENTS IN THE WORLD COPPER SECTOR                  02/16/84    14.28.26.    PAGE    29
            REFERENCE MAP OF VARIABLES


VARIABLES   TYPE    REFERENCES

XFD         PARAM            REF      822      823      824      825      826      827      852      862      872
                             883  DEFINED      820      821      822      823      824      825      826      DCL
                             801
XFR         VAR              REF      707      737      742      769      771      820      DCL      682
XFS         VAR              REF      707      740      742      768      772      821      DCL      683
XI          VAR              REF      707      736      737      768      810      811      812      DCL      680
XII         PARAM            REF      814      815      816      817      818      827      834      843      852
                         DEFINED      810      811      812      813      814      815      816      817      818
                             DCL      800
XIR         VAR              REF      707      737      739      769      771      813      DCL      681
YBASE       PARAM            REF    2*130      133      451  DEFINED      129      DCL      125
YM          VAR              REF      708      748      752      DCL      693
YS          VAR              REF      708      750      754      DCL      694
ZM          VAR              REF      707      736      744      756      764      811      812      833      835
                             836      842      844      845      851      853      854  DEFINED      784      DCL
                             678
ZS          VAR              REF      707      739      746      766      861      863      864      871      873
                             874      882      884      886  DEFINED      785      DCL      679

SETS

  C           COMMODITIES
  CF          FINAL PRODUCTS
  CFR         FINAL PRODUCTS FROM PROCESSING
  CFS         FINAL PRODUCTS FROM SEMI-MANUFACTURE
  CGR         CONSUMPTION GROWTH RATE INTERVALS
  CIL         SCRAP TYPES
  CIM         FINAL PRODUCTS AT MINES AND SMELTERS
  CM          COMMODITIES IN MINING AND PROCESSING
  CPS
  CS          COMMODITIES AT WIRE TUBE & SHEET PLANTS
  I           MINE SMELTER AND REFINERY LOCATION
  IH          REFINERIES & SMELTERS WITH HIGH COST
  IIT         MAP OF LOCATIONS TO CLASSIFICATION
  IL          REFINERIES & SMELTERS WITH LOW COST
  IP          ALIAS FOR I
  IRUN        MINE SMELTER AND REFINERY LOCATIONS
  IRUNP       ALIAS FOR IRUN
  IT          COPPER PROCESSING CLASSIFICTION
  J           WIRE TUBE AND SHEET PLANT AND MARKET LOCATIONS
  JH          SEMI-MANUFACTURING LOCATIONS WITH HIGH COST
  JL          SEMI-MANUFACTURING LOCATIONS WITH LOW COST
  JP          ALIAS FOR J
  JRUN        WIRE TUBE AND SHEET PLANT AND MARKET LOCATIONS
  JRUNP       ALIAS FOR JRUN
  JT          SEMI-MANUFACTURING CLASSIFICATION
  JTJ         MAP OF LOCATIONS TO CLASSIFICATION
  M           PRODUCTIVE UNITS
  MGR         MAP OF INTERVAL TO YEARS
  MM          PRODUCTIVE UNITS AT MINING AND PROCESSING PLANTS
  MS          PRODUCTIVE UNITS AT SEMI-MANUFACTURE PLANTS
  N           OF NODES - ALL LOCATIONS
  NP          ALIAS FOR N
  NPP         ALIAS FOR N
```

```
GAMS 1.208  MODELING INVESTMENTS IN THE WORLD COPPER SECTOR                          02/16/84    14.28.26.   PAGE   30
            REFERENCE MAP OF VARIABLES


SETS

  NTFR        NO TARIFFS ON REFINED COPPER BETWEEN
  P           PROCESSES
  PM          PROCESS AT MINES SMELTERS AND REFINERIES
  PMH         HIGH GRADE ORE MINING PROCESSES
  PMM         MINING PROCESSES
  PR          REFINING PROCESSES
  PS          SMELTING PROCESSES
  PSH         SHEET PLANT PROCESSES
  PSI         REFINERY AND SMELTER PROCESSES USING SCRAP COPPER INPUT
  PSJ         WIRE TUBE AND SHEET PROCESSES USING SCRAP COPPER INPUT
  PSM         SEMI-MANUFACTURING PROCESS
  PTR         TUBE AND ROD PLANT PROCESSES
  PW          WIRE PLANT PROCESSES
  T           SOLUTION TIME PERIOD
  TBASE
  TG          TIME PERIODS
  TGP         ALIAS FOR TG
  TG1         INTERVAL 1 DEFINITION
  TG2         INTERVAL 2 DEFINITION
  TP          ALIAS FOR T

PARAMETERS

  A           INPUT-OUTPUT COEFFICIENTS
  B           CAPACITY UTILIZATION MATRIX
  CAPM        ESTIMATES OF MINE SMELTER AND REFINERY CAPACITIES IN 1980 (1000 TONS)
  CAPS        EXISTING CAPACITY IN SEMI-MANUFACTURING IN 1980 (1000 TONS)
  COSTT       COST TABULATION (MILLION US$)
  CPDM        CAPACITY-PRODUCTION AT MINES          (1000 TONS)
  CPDR        CAPACTIY-PRODUCTION AT REFINERIES     (1000 TONS)
  CPDS        CAPACITY-PRODUCTION AT SMELTERS       (1000 TONS)
  CPDSH       CAPACITY-PRODUCTION AT SHEET PLANTS   (1000 TONS)
  CPDT        CAPACITY-PRODUCTION AT TUBE PLANTS    (1000 TONS)
  CPDW        CAPACITY-PRODUCTION AT WIRE PLANTS    (1000 TONS)
  DEMAND      DEMAND FOR REFINED COPPER AND SEMI-MANUFACTURES (1000 TONS)
  DEM1980     CONSUMPTION IN 1980 AND GROWTH RATES (1000 TONS & %): TABLE
  DIS         DISCOUNT FACTOR
  DISTANCE    SEA DISTANCES IN NAUTICAL MILES
  DRAIL       RAIL DISTANCES (KM)
  HBARM       ECONOMIES OF SCALE SIZE FOR PROCESSING PLANTS (1000 TONS)
  HBARS       ECONOMIES OF SCALE SIZE FOR SEMI-MANUF PLANTS (1000 TONS)
  HDS1        SCRAP USAGE IN 1980 AT SMELTERS AND REFINERIES (1000 TONS)
  HDS2        SCRAP USAGE IN SEMI-MANUFACTURING IN 1980 (1000 TONS)
  HHATM       MAXIMUM SIZES FOR PROCESSING PLANTS              (1000 TONS)
  HHATS       MAXIMUM SIZES FOR SEMI-MANUFACTURING PLANTS      (1000 TONS)
  INVM        INVESTMENT DATA FOR MINES - 1980(1)
  INVRS       INVESTMENT COSTS FOR REFINERY AND SMELTER - 1980(1)
  INVS        INVESTMENT COSTS FOR SEMI-MANUFACTURING - 1980
  ITP         IMPORT TARIFFS (%)
  LBPERTON    CONVERSION FROM LBS TO TONS
  LIFE        LIFE OF UNITS   (YRS)
  MAPIC       COMMODITIES AT MINES REFINERIES AND SMELTERS
  MAPIP       PROCESSES AT MINES REFINERIES AND SMELTERS
  MAPJC       COMMODITIES AT WIRE TUBE AND SHEET PLANTS
```

```
GAMS 1.208  MODELING INVESTMENTS IN THE WORLD COPPER SECTOR                          02/16/84    14.28.26.   PAGE   31
            REFERENCE MAP OF VARIABLES


PARAMETERS

  MAPJP       PROCESSES AR WIRE TUBE AND SHEET PLANTS
  MUFS        TRANSPORT COST: FINAL: SEMI-MANUFACTURES FROM LOCATIONS TO MARKETS (US$ PER TON)
  MUI         TRANSPORT COST: REFINED COPPER TO SEMI-MANUFACTURE AND MARKETS     (US$ PER TON)
  MUR         TRANSPORT COST: RAW MATERIAL & INTERMEDIATE GOODS                  (US$ PER TON)
  NUM         PROPORTIONAL CAPITAL COST OF MINING AND PROCESSING PLANTS (MILLION US$ PER 1000 TPY)
  NUS         PROPORTIONAL CAPITAL COST OF SEMI-MANUF PLANTS            (MILLION US$ PER 1000 TPY)
  OM          MINE OPERATING COST IN 1980(1)  (US$ PER TON OF COPPER CONTENT)
  OMEGAM      SCALE COST OF MINING AND PROCESSING PLANTS                (MILLION US$ PER 1000 TPY)
  OMEGAS      SCALE COST OF OF SEMI-MANUACTURING PLANTS                 (MILLION US$ PER 1000 TPY)
  OPI         OPERATING COST ESCALATOR FOR MINING AND PROCESSING LOCATIONS
  OPJ         OPERATING COST ESCALATOR FOR SEMI-MANUFACTURING LOCATIONS
  OPM         MINING AND PROCESSING COSTS IN 1980(1) (US$ PER TON OF COPPER CONTENT)
  OPR         REFINERY OPERATING COST IN 1980(1) (US$ PER TON OF COPPER CONTENT)
  OPS         OPERATING COSTS FOR SEMI-MANUFACTURING IN 1980(1) (1000 US$ PER TON)
  OS          SMELTER OPERATING COST IN 1980(1) (US$ PER TON OF COPPER CONTENT)
  PC          COMMODITY PRICES (US$ PER TON)
  PERIOD      INTERVAL LENGTH (USED IN OBJECTIVE FUNCTION)
  RESC        ANNUAL EXTRACTION LIMITS ON MEDIUM AND HIGH GRADE ORES (THOUSAND TONS)
  RESERVES    ORE RESERVES ESITMATES FOR NEXT 20 YEARS (1980-81 - MILLION TONS)
  RHO         DISCOUNT RATE
  RPH         LENGTH OF PERIODS FOR RESERVES
  SCRAPI      SCRAP AVAILABILITY AT REFINERIES AND SMELTERS (1000 TONS)
  SCRAPJ      SCRAP AVAILABILITY AT SEMI-MANUFACTURERS      (1000 TONS)
  SFI         SITE FACTOR FOR MINING AND PROCESSING PLANTS
```

```
  SFJ        SITE FACTOR FOR SEMI-MANUFACTURING PLANTS
  SIGMA      CAPITAL RECOVERY FACTOR
  SPR        SCAP PRICES (AT COPPER PRICE OF US$2000 A TON)
  SPRICE     SHADOW PRICE OF DEMAND BALANCE
  TARIFFR    TARIFFS ON REFINED COPPER  (US$ PER TON OF COPPER CONTENT)
  TARIFFS    TARIFFS ON SEMI-MANUFACTURED GOODS (US$ PER TON  OF COPPER CONTENT)
  TS         TIME SUMMATION MATRIX
  UC         TRANSPORT COST OF SEMI-MANUFACTURES AND ALL OTHER TYPES OF INTERMEDIATES
  XFD        SHIPMENTS OF FINAL PRODUCTS: DISAGGREGATED       (1000 TONS)
  XII        SHIPMENTS OF INPUTS (ORE-BLISTER-REFINED COPPER - 1000 TONS)
  YBASE      DISTANCE OF YEAR FROM BASE YEAR

VARIABLES

  HM         CAPACITY EXPANSION: MINES SMELTERS AND REFINERIES                (1000 TPY)
  HS         CAPACITY EXPANSION: WIRE TUBE AND SHEET PLANTS                   (1000 TPY)
  PHIKM      COSTS: CAPITAL CHARGES AT MINES SMELTERS AND REFINERIES       (MILLION US$)
  PHIKS      COSTS: CAPITAL CHARGES AT WIRE TUBE AND SHEET PLANTS          (MILLION US$)
  PHIOM      COSTS: OPERATING COSTS AT MINES SMELTERS AND REFINERIES       (MILLION US$)
  PHIOS      COSTS: OPERATING COSTS AT WIRE TUBE AND SHEET PLANTS          (MILLION US$)
  PHIT       COSTS: TRANSPORT COSTS                                        (MILLION US$)
  PHITF      COSTS: TARIFF COSTS                                           (MILLION US$)
  PHIU       COSTS: TOTAL ANNUAL UNDISCOUNTED COST                         (MILLION US$)
  PHIUTF     COSTS: TOTAL ANNUAL UNDISCOUNTED COST WITH TARIFFS            (MILLION US$)
  PHI1       TOTAL COST                                                    (MILLION US$)
  PHI2       TOTAL COST WITH TARIFFS                                       (MILLION US$)
  SM         UNUSED ECONOMIES-OF-SCALE EXPANSION: MINES SMELTERS AND REFINERIES  (1000 TPY)
  SS         UNUSED ECONOMIES-OF-SCALE EXPANSION: WIRE TUBE AND SHEET PLANTS     (1000 TPY)
  SSA        SCRAP SUPPLY: SEMI-MANUFACTURING                                 (1000 TPY)
  SSM        SCRAP SUPPLY: SHEET AND TUBE PLANTS                              (1000 TPY)
```

```
GAMS 1.208  MODELING INVESTMENTS IN THE WORLD COPPER SECTOR                    02/16/84    14.28.26.   PAGE   32
            REFERENCE MAP OF VARIABLES


VARIABLES

  SSR        SCRAP SUPPLY: SMELTERS AND REFINERIES                            (1000 TPY)
  XFR        SHIPMENTS: REFINED COPPER TO MARKETS FOR END USE                 (1000 TPY)
  XFS        SHIPMENTS: SEMI-MANUFACTURES TO MARKETS                          (1000 TPY)
  XI         INTERPLANT SHIPMENTS OF ORE AND BLISTER                          (1000 TPY)
  XIR        SHIPMENTS: REFINED COPPER FROM SMELTERS TO SEMI-MANUFACTURERS    (1000 TPY)
  YM         EXPANSION DECISION VARIABLE: MINES SMELTERS AND REFINERIES
  YS         EXPANSION DECISION VARIABLE: WIRE TUBE AND SHEETS PLANTS
  ZM         PROCESS LEVEL: MINES SMELTERS AND REFINERIES                     (1000 TPY)
  ZS         PROCESS LEVEL: WIRE SHEET AND TUBE PLANTS                        (1000 TPY)

EQUATIONS

  AKM        ACCOUNTING: CAPITAL CHARGES FOR MINES SMELTERS AND REFINERIES  (MILLION US$)
  AKS        ACCOUNTING: CAPITAL CHARGES FOR WIRE TUBE AND SHEET PLANTS     (MILLION US$)
  AOBJ       OBJECTIVE FUNCTION                                             (MILLION US$)
  AOBJTF     OBJECTIVE FUNCTION WITH TARIFFS                                (MILLION US$)
  AOM        ACCOUNTING: OPERATING COST FOR MINES SMELTERS AND REFINERIES   (MILLION US$)
  AOS        ACCOUNTING: OPEARTING COST FOR WIRE TUBE AND SHEET PLANTS      (MILLION US$)
  AOT        ACCOUNTING: TRANSPORT                                          (MILLION US$)
  AOTF       ACCOUNTING: TARIFFS                                            (MILLION US$)
  AU         ACCOUNTING: UNDISCOUNTED ANNUAL COST                           (MILLION US$)
  AUTF       ACCOUNTING: UNDISCOUNTED ANNUAL COST WITH TARIFSF              (MILLION US$)
  CCM        CAPACITY CONSTRAINT: MINES SMELTERS AND REFINERIES                (1000 TPY)
  CCS        CAPACITY CONSTRAINT: SEMI-MANUFACTURING                           (1000 TPY)
  ICM1       MAXIMUM EXPANSION: MINES SMELTERS AND REFINERIES                  (1000 TPY)
  ICM2       LIMITS TO ECONOMIES-OF-SCALE: MINES SMELTERS AND REFINERIES       (1000 TPY)
  ICS1       MAXIMUM EXPANSION: WIRE TUBE AND SHEET PLANTS                     (1000 TPY)
  ICS2       LIMITS TO ECONOMIES-OF-SCALE: WIRE TUBE AND SHEET PLANTS          (1000 TPY)
  MBM        MATERIAL BALANCE: MINES SMELTERS AND REFINERIES                   (1000 TPY)
  MBS        MATERIAL BALANCE: SEMI-MANUFACTURING                              (1000 TPY)
  MR         MARKET REQUIREMENTS                                               (1000 TPY)
  OREC       HIGH-GRADE ORE MINING LIMITATIONS                                 (1000 TPY)
  SBS        SCRAP BALANCE AT SEMI-MANUFACTURING LOCATIONS                     (1000 TPY)

MODELS

  COPPER
```